D1549792

"Parry has deftly removed egotistical consciousness from ... -
tery of things and reinserted it integrally back into nature, and in doing
so everything—thought, time, space, humanity—comes to look different,
more radiant, at once older and newer. [This] is even more than a prescrip-
tion for restoring the world.... In the pages of *Original Thinking* the world is
already healthy once more."—JEFFREY PAINE, author of *Re-Enchantment*

"For over a decade, Glenn Aparicio Parry has been convening a unique
gathering of minds, bringing a distinguished array of indigenous wisdom
keepers together with Western scientists—not to hold forth in front of an
audience but simply to be in rich dialogue with one another, listening,
learning, discovering fresh ways of thinking and understanding our aching
world."—DAVID ABRAM, author of *The Spell of the Sensuous*

"The heart of Parry's journey taps the living pulse of sacred, deep dialogue
and calls upon the reader to become an awakened, 'original' reader, entering
a sacred rite of passage toward the whole Self, heart and soul."—ASHOK
GANGADEAN, PhD, author of *Awakening the Global Mind*

"Parry's gift is to bring out the core essence of humanity, love, and prayer as
well as our infinite connection to nature."—HARLAN MCKOSATO (Sac
and Fox), former host of *Native America Calling*

"*Original Thinking* is a call for humanity to return to our original purpose
in the cosmos. Asking the most basic questions, Parry challenges us to set
aside conventional thinking and ponder life's firsts as blessings for which to
be thankful—the first thought, the first awareness, the first consciousness,
the beginning of the cosmos—in order to reeducate our minds, become one
with nature as a state of being, and achieve wholeness through Spirit as our
source."—PHILLIP H. DURAN (Tigua), author of *The Condor and the Eagle*

"Read this book and be liberated from the selfish, greedy, egotistic, and
separatist thinking that is currently dominating and destroying our planet.
Learn how to return to the origin of thought which brings us back into
synchronization with Nature. It isn't hard. Parry provides excellent instructions."—CAROLE HART, producer and director of *For the Next Seven Generations: 13 Indigenous Grandmothers Weaving a World that Works*

"A brave and bold book that draws upon the power of ancient wisdom, ways of being in the world still rooted to a vibrant sense of the 'whole'—the whole of nature, the whole of humanity, the whole of time. Here is a book that we actually *need* to read, for our sake and the sake of life on this planet. For those willing to risk thinking their way toward viable and necessary alternatives to the dominant pressures of globalized capitalism, for those aware of the drastic consequences of our ignorance of the spiritual in favor of a selfish materialism, this book is not simply a treasure. It may be the one guide we need to begin to sort out how we might yet save ourselves."—MARK S. BURROWS, coauthor of *Minding the Spirit*

"*Original Thinking* has to do with the movement between civilizations, between cultures, between languages, between worlds. It has to do with overcoming boundaries, crossing borders not only geographically—but also psychologically and spiritually—in communication, in thinking, and in identity. It means that one is never finished, that one is always journeying onward. That one is always ready for the next experience."—MONA POLACCA (Havasupai/Hopi/Tewa), member of International Council of Thirteen Indigenous Grandmothers

"There is a lot of rich terrain to explore in the 'third space' opened by the decade-long dialogue between indigenous elders and scientists that inspired this work. As one of the long-time participants who was personally transformed, I am delighted by how the book distills some of the wisdom and burning questions that were forged in our historic dialogues. And I am even more heartened by Parry's ability to translate that wisdom into practical solutions and actions."—MATTHEW C. BRONSON, PhD, coeditor of *So What? Now What? The Anthropology of Consciousness Responds to a World in Crisis*

"Reading *Original Thinking* felt like being stirred and mixed and rocked in a cradle between my left brain and my right brain. And I do suspect that there was an interconnecting subtle weaving process stitching together an even more supple fabric within my whole brain. And that this fabric that re-wove me into it re-wove me into the original memory of Wholeness of all of Creation."—VICTORIA HANCHIN, author of *The Seer and The Sayer*

ORIGINAL THINKING

A Radical ReVisioning of Time, Humanity, and Nature

GLENN APARICIO PARRY

Foreword by James O'Dea

North Atlantic Books
Berkeley, California

Published by

North Atlantic Books   Cover art: "Milky Way" by James Jereb

Berkeley, California   Cover and book design by Suzanne Albertson

Printed in the United States of America

Original Thinking: A Radical ReVisioning of Time, Humanity, and Nature is sponsored and published by the Society for the Study of Native Arts and Sciences (dba North Atlantic Books), an educational nonprofit based in Berkeley, California, that collaborates with partners to develop cross-cultural perspectives, nurture holistic views of art, science, the humanities, and healing, and seed personal and global transformation by publishing work on the relationship of body, spirit, and nature.

North Atlantic Books' publications are available through most bookstores. For further information, visit our website at www.northatlanticbooks.com or call 800-733-3000.

Library of Congress Cataloging-in-Publication Data

Parry, Glenn Aparicio, 1955–

    Original thinking : a radical revisioning of time, humanity, and nature /

Glenn Aparicio Parry.

      pages cm

    Summary: "Glenn Aparicio Parry organized and participated in thirteen ground-breaking dialogues between Native American elders and leading-edge Western scientists that explored the underlying principles of the cosmos. Inspired by these dialogues, *Original Thinking* unfolds in a similar way to a dialogue circle. The questions it asks penetrate ever deepening layers of meaning, such as, Is it possible to come up with an original thought?, What does it mean to be human?, and How has our thinking created our world today? Delving into the evolution of Western thought, Parry seeks to recover the living roots of wisdom that can correct the imbalances in our modern worldview."—Provided by publisher.

    ISBN 978-1-58394-890-3 (pbk.)—ISBN 978-1-58394-891-0 (ebook)

    1. Thought and thinking. I. Title.

    BF441.P3747 2015

    191.089'97—dc23   2014036414

1 2 3 4 5 6 UNITED 19 18 17 16

Printed on recycled paper

OPENING PRAYER

Beloved Creator, Great Mystery, Great Spirit that underlies all things I see, touch, taste, hear, and smell … On this beautiful morning,

To the *East*, I make an offering to the Sun, place of Origin, where my life and my thoughts come from, and I thank you for the amazing gift of life for which I will be grateful for all my days to come and I promise to live my life well and serve the people and all our relations.

To the *South*, I thank the elements, the Light, the Air, the Water, and the Earth. You are what makes up creation and I give thanks to you because I too am made of you and I pray that you be returned to your original state of purity and wholeness. I pray that all the people understand that we are made of the elements, that you not only make up life but *are life*, *are awareness*, and that you be treated with the respect and reverence you deserve.

To the *West*, I make an offering to the ancestors, to all the people of the spirit world that have been and still are teachers to me: Grandfather Tobasonakwut Kinew, Grandfather Leon Secatero, Grandfather Dan Moonhawk Alford, Grandmother Paula Gunn Allen, Grandfather Bohm, Grandfather Krishnamurti, Grandmother Blavatsky, and my mother and father, Joan Parry and Juan Paricio Parry.

To the *North*, I thank all the four-leggeds, the deer, the elk, the bobcat, fox, and especially the bear for his introspective wisdom and medicine. I thank the flying ones, the songbirds, finches, and jays who serenade me daily, the quail, the owls who visit in sight or sounds. And special thanks for the soaring birds of prey, the hawks and eagles for their gift of vision and oversight. I thank the finned ones who swim in the rivers and oceans, and I pray for the finned ones that their waters be clean again; I pray for the crawling ones and the burrowing ones who make space in the earth for next year's plantings, the snakes who teach us to shed our infrastructures and embrace change, and for the change in the seasons that teach us the timing of Mother Earth.

I thank you, Beloved Creator, for all the immense beauty that is everywhere—in the mountains, the valleys, the rivers, the oceans, the mesas, buttes, and caves, with special gratitude for the plants and trees who are the lungs of Mother Earth. I am grateful for every breath I take and consciously offer my breath back to the plants and the trees, thanking you for the beautiful partnership we share, breath for breath, for your special medicine, your nutrition, your companionship. I thank the clouds that form over the mountains bringing rain and snow to the thirsty valley that sustains the fruits and vegetables that feed me in body and spirit as well as all the other hungry mouths that need them. I thank my wife, my four-legged companions, and the plants who share my indoor space for their love and support, and also the unseen helpers, including the bacteria in my own body, and the living presence of wood from my old drafting table that serves as my writing desk, the bamboo floor it rests upon, and the trees I am blessed to see from my window: the aspens, the piñons, the Ponderosa pines as well as the cottonwoods down by the river; and I pray that the river be returned to her full health and that all rivers be undammed and allowed to flow free as is their nature.

Beloved Creator, I thank you for the blessing of being able to share what I have come to experience and even though my knowledge will never be complete, I pray that I do not hold onto knowledge but allow it to come forward as needed in service to whomever needs it and I am grateful for you that feeds me from the original source and I trust this wellspring to be always whole, complete, and timeless.

I pray that the message I share benefits not just humankind, but the many relatives I share this beautiful planet with. I pray to and through the tobacco within my pipe and send the prayers to the heavens and to all the realms I cannot travel to so that my prayer is heard by all that need to hear. Aho. Mitakuyasen (May it be so. I say these things for all my relations).

To Tomoko, with love

CONTENTS

FOREWORD
by James O'Dea

For the last thirty years the Kogi people have been sending urgent messages to the world from atop Colombia's Sierra Nevada de Santa Marta, the world's highest coastal mountain range. They view themselves as elder brothers and sisters whose younger siblings across the planet have fallen out of rhythm with the matrix that nourishes all life: we have stopped listening to the heartbeat of Mother Earth. We are no longer pulsed by Earth's abundant generosity towards all life. Instead we are fuelling a hyper-acceleration we call progress, racing beyond the laws of ecology and sustainability to a paradise defined by the acquisition of stuff. Ignoring Nature's principles for sustaining biodiversity and replacing them with manufactured goods and techno wizardry is leading to a world where everything begins to look monochromatic and is dangerously out of balance. The Kogi do not see this outright assault on Nature in support of materialist progress as representing a final Armageddon. Even as they predict the possibility for catastrophic loss of life resulting from ignoring Nature's warnings about the real cost of our lifestyles, they defer to the original intelligence of Nature as the guardian of all life. Kogi elders personally told me, "Mother Nature knows how to clean herself and how to heal herself. Humanity must understand that Nature has infinite patience to teach us how to live in harmony with her laws."

The Kogi people have a commitment to the whole planet and daily pray for the world's people, its mountains, its rivers, its forests, and oceans. Their elders are in constant telepathic communion with life on Earth. They are a deeply wisdom-sourced culture. Yet they are not immune to the disruptions to their way of living brought about by climate change, aggressive commercialism, political turmoil, and the seductions of consumerist society. Their culture has remained relatively intact compared to some other Native communities, but the velocity of social and cultural destruction that comes with "progress" has already begun to affect them.

The wise Bushmen of the Kalahari have a story deeply relevant to this era. They speak of the time of All Devourer. This seems a particularly apt archetype to represent the current rapacious destruction of eco-systems and natural resources. All Devourer eats up everything, including rivers and trees. But like the Kogi, the original people of the Kalahari do not interpret these calamities as foreshadowing an end to the world; rather they provide a basis for understanding cycles and processes. Since what it has swallowed is living, it only makes sense to the Bushmen that Life itself will eventually emerge from the belly of All Devourer. A time of renewal will follow a time of destruction and the wise are the ones who know what their roles and responsibilities are in each phase.

We are indeed in the time of All Devourer. For this reason it is particularly opportune that Glenn Aparicio Parry offers us the exceptional creativity and wisdom contained in *Original Thinking*. Rather than fixate on the breakdown we see everywhere, Parry chooses to go beyond symptoms to root causes. He enters into the prevailing maps of the mind with enormous precision. He shows us the trajectory of the predominant western worldview as it hurtles towards alienation and fragmentation. He unravels our predicament of creating a world where we serve accelerating time rather than the rhythms of a deeper remembrance of who it is we really are.

A spirit of resurrection and renewal pervades the book as Parry revisions a world that does not jettison the past but fully integrates its learning into a richly sensorial experience of harmony and wholeness. He achieves what so many others fail to do in honoring the past—he does not get stuck in looking back but gives us a template of meaning for living human values that are integral to the natural order rather than being in rebellion against it. *Original Thinking* is a hymn to the recovery of being in a mechanized world and to our liberation from paradigms that severely inhibit our innate wisdom.

The bedrock of experience for Parry's own fertile imagination and remarkable insightfulness emerges from years of dialogic practice. Dialogue has also been a path for me in my own journey to explore our capacities for collective social healing. I have facilitated and witnessed the healing power of dialogue between protagonists in major conflicts around the world, between victims and perpetrators, with survivors of torture, with human rights activists, peace builders, psychologists, and academics. Dialogue can create a field of openness to listening and being with another that can be profoundly healing. In dialogue we can move beyond linear narratives and constant disagreements over facts and history. We can move into an exploration of the truth of experience. Invariably the field of connectivity between people becomes significantly deeper when subjective truth is honored and the wisdom of the heart is not suppressed.

As you will read in the pages ahead Parry has a great deal to share on the kind of integral dialogue that gives full play to both head and heart: to feeling with the mind and thinking inside the heart. But our contemporary systems are not organized around dialogic process. Our political processes are polarized. Our educational systems are compartmentalized and structured around information transmission, largely discounting the relevance of subjective experience. Dialogue is not seen as having the muscle needed to generate breakthroughs, originality, creativity, or real momentum in the social order. From this perspective, dialogue is a kind of tea-time affair that may be emotionally satisfying but not impactful on civilization as a whole.

You are about to discover an approach to dialogue that is transformational for our collective engagement with Nature and with a thriving planetary civilization. Our deepest thinking is relational and it communes with all life rather than dissecting it into separate units. *Original Thinking* invites you to re-experience your own ground of being and your relationship to Cosmos, not as an information retrieval process but as a profound journey of discovery sourced by an open and dialogic inquiry into the connection between you and all that is. What a quest!

Finally a map that drops you into a territory of meaning where once more you have to trust your instincts, hone your intuition, and let your inner compass guide you to those wisdom keepers who will have the perfect question to set you on your way. I assure you some paradigm shifting, and even life-changing, questions await you in this important book.

James O'Dea,

author of *The Conscious Activist: Where Activism Meets Mysticism*

PREFACE

This book emerges out of the Language of Spirit dialogue conferences, sponsored by the SEED Institute, an educational institution I founded. These seminal gatherings were held annually in Albuquerque, New Mexico, USA, between 1999 and 2011, and brought Western scientists together with Indigenous elders to explore the underlying reality of the cosmos.

It took an unusual confluence of events for me to come to appreciate dialogue, for it is a practice that often feels alien to the Western mind. The first occurred when I took a graduate school class in anthropological linguistics taught by the late Dan Moonhawk Alford at the California Institute of Integral Studies in 1983. Moonhawk, who once described himself as "standing at the lonely intersection of consciousness, linguistics, Native America, and quantum physics," included all of the above in his class. This was a new confluence of ideas for me, and probably for all my fellow students. In fact, Moonhawk was so confident this class would change our lives that he assigned a mandatory diary to chart the changes in our thinking. He was right; I would never be the same again. As a person, Moonhawk was one of the most welcoming and warm individuals I have met. I became lifelong friends with him and his teaching assistant at the time, Matthew Bronson, who later became an associate professor at the California Institute of Integral Studies.

About ten years later, Moonhawk was invited to participate in a unique and historic series of dialogues. The inspiration for the dialogues came from the former director of Native Studies at Harvard University, Leroy Little Bear, who had been quietly observing that quantum theory was beginning to bring the Western science worldview into alignment with the Blackfoot worldview. Both cutting edge quantum theory and the traditional Blackfoot worldview saw everything in the cosmos as being in dynamic flux. Little Bear also sensed that David Bohm, a brilliant, visionary physicist, would be more open to listening to a Native viewpoint than most Western scientists, as Bohm had an appreciation for dialogue that was similar to Native talking circles.[1]

Bohm, for his part, understood that the English language is a barrier to comprehending reality as an interconnected, dynamic whole because it is so noun-oriented. What was needed, Bohm felt, was a much more verb-oriented language, and he had been experimenting with creating a new mode of language that he called the "rheomode," from the Greek *rheo,* to flow. This too was a sign, for Little Bear knew that the Blackfoot language operated in a way that was very similar to the form of language Bohm was trying to create. And so, together with David Peat, the author of *Blackfoot Physics,* Little Bear approached Bohm hoping to initiate a meeting of the minds between Western scientists and linguists and Indigenous scholars and elders.

The first of these monumental gatherings took place in 1992 at the Fetzer Institute in Kalamazoo, Michigan—exactly five hundred years after Columbus set foot on this continent. It was truly a historic event: the first time Indigenous elders and Western scientists met on equal footing in the postcolonial era. Although Bohm died shortly after the first dialogue, something had been set in motion, and a succession of groundbreaking dialogues were held on an annual or semiannual basis, mostly in Banff, Canada.

Then, one day, Moonhawk and I were sitting outside in my backyard overlooking the Sandia Mountains. He was smoking his pipe, as he often did, and out came a pipe dream. He turned to me and said, "What if we bring the dialogues to Albuquerque?" He meant, what if SEED sponsored the dialogues?

This was a bit out of the blue, coming only two years after I had founded SEED as an educational nonprofit. But as soon as he suggested it, I knew it would happen. I said to him, "Yes! Let's do it, and if we can't get a grant, let's bring an audience to listen." (It is, after all, a principle of quantum theory that the observer affects the observed.) So, the very next year, in 1999, SEED began an annual tradition of hosting what became known informally as the Southern Circle, and publicly as the Language of Spirit dialogues. Some form of dialogue has continued ever since, even as the Language of Spirit dialogues were retired in 2011 after thirteen years.

In that first year, Moonhawk selected all the participants for the dialogue circle, except one. He was not familiar with the elders of New Mexico, so he asked me to select one elder who would be most suitable for a dialogue on Indigenous and Western sciences. I took this responsibility seriously and began a thorough search for the right person. Eventually, I was led to the late Headman of the Canoncito Band of Navajo, Grandfather Leon Secatero. (I thank Terry Fisher, Patricio Dominguez, and Cynthia Walker, among others, for leading me to this wise old soul.) We were blessed to have many wonderful individuals participate in the dialogues, but of them all, Grandfather Leon was the most peaceful and harmonizing influence. When Grandfather Leon spoke, his voice was so gentle and mellifluous it was as if he were singing you a lullaby. He taught me many things that are still growing like a seed inside of me—because when he spoke in the way that he did, he was carrying in his voice a form of sacred knowledge that could not be carried by words alone.

These dialogues were meaningful and even transformational for many people, and certainly for me. But superlatives cannot convey the feeling of what transpired in those special gatherings, so let me try to paint the picture.

Imagine an inner circle of people with an opening to the east, surrounded by an outer circle in the same configuration. All the people in the room, both inner and outer circles, are engaged in considering the same question, consciously thinking together and listening carefully to what is being said. Gradually, as if the moderator is tuning up an orchestra, we begin to harmonize. There are many stories shared, and some humor sprinkled in. The flow of the conversation is at times intellectual, then emotional, even spiritual, and often culminates in deep healing. Imagine thinking of something to say and, at that very moment, witnessing someone else opening their mouth and saying exactly what was on your mind.

This happens again and again, until suddenly you realize that thought is alive and moving through the room—that the dialogue is being informed by a higher consciousness.

What you are picturing is a well-run dialogue circle—something I had the privilege of participating in for the past fifteen years under the expert moderation of Little Bear. It is because of these experiences that I was able to feel the power of thought; to see the possibility of thinking together as one mind; to understand that thought, held in a sacred context, can be a prayer. It is through this unique modern form of participatory consciousness—which was both a group construction of knowledge and the invocation of wisdom—that the seeds of this book began to form.

I am eternally grateful to Leroy Little Bear for moderating those dialogues, and to everyone that participated in them, both inner and outer circles. It is not an overstatement to say that during the past fifteen years, these dialogues—and dialogue itself—have changed my life. I have learned, and am still learning, to slow down my thinking processes in order to allow a deeper knowing to come through. I am learning that the most important thing we can do is to listen—to each other and to the natural rhythms that surround us.

I should make clear, however, that this book does not necessarily represent anyone else's point of view in the dialogue circle. It is limited to what I took away from these experiences, filtered through my own ears and eyes, and no doubt through my own prejudices and preconceptions; for try as we might to set aside these biases, they remain. In short, this book represents my own unique journey through the dialogues and beyond. It is my own interpretation of what was said and how that has entered my life to produce the understanding I now have. It is also informed by many other life experiences, including Native American ceremonies I have been blessed to participate in, and my own direct experiences in nature.

I do not claim to be an expert on Native America. I may have learned a great deal from my Native American friends, but there is a whole lot more that I do not know. In this, I follow Moonhawk, who was my mentor into Native America. Moonhawk often told a certain story when he wished to illustrate the depth of Native American wisdom, a story from his childhood. When Moonhawk was about ten years old, he visited his father at his place of work. His father was a longshoreman, and they spent the morning together on a jitney, stacking up logs and other things, and that was great fun for a boy. Then, lunchtime came. Moonhawk's father was working on the California coast amongst the tall redwoods, and they took their lunch at the edge of a cliff overlooking the Pacific Ocean. This was the first time Moonhawk had ever seen the ocean, and he remarked, "Dad, that sure is a lot of water out there." His father slowly looked out over the vast expanse, and replied, "Yep, and that's just the top of it."

INTRODUCTION

In hearing the phrase "original thinking," you probably anticipate something new and innovative—and you may get it. But the title of this book is intentionally playful. This book is about thinking—the origin and full continuum of thinking—in the past and in the now.

Thought is vibration. Thoughts of the past may dissipate, but they never quite disappear. They continue to matter in the present. The web of thought is vast and mysterious, and unfolds at its own pace.

The sum total of thought is tantamount to the *intelligence* of nature. Nature is intelligent in part because it contains volumes of stories. These stories reside in particular places—in rocks, caves, rivers, and oceans; but they also penetrate the memories, dreams, and reflections of all creatures. These stories are original and immutable. They are part of the fabric of life.

Our story is both new and old because it is connected to the origin of all stories. It includes humans but is not exclusively about humanity. Nothing ever is.

A very long time ago, people first gathered around a fire to tell stories. Storytelling is the oldest form of communication known to humankind. Today, people continue to gather in this way, and it has a name: *talking circle,* also known as dialogue. Talking circle is a model of inclusiveness, for all voices are welcome, and everyone is able to see each other's faces. Talking circle is the origin of what became known as consensus-making, where decisions are made as a group.

But there is more to talking circle than meets the eye. These kinds of circles typically have an opening in one direction, which was originally a way of inviting spirit into the circle from outside the tribe. The opening allows the spirit of nature to enter and speak directly through the members.

This was because, unlike today, we once believed our thinking emerged directly from nature. People were understood to be vessels—conduits for thought—not the originators of our thoughts. To think was to receive inspiration from nature. Thoughts emerged from Great Spirit, or the womb of creation—moved through us, and then returned to the Great Mystery. *Thought and Spirit were the same thing.* All of creation was once understood as a dance of the elements: fire, air, water, and earth. Everything—not just humans—was known to be imbued with spirit—with consciousness. Our thoughts were sacred because they connected us to All there is.

The origin of the word "thinking" is *thanking*[1] because our thoughts were originally prayers. The original purpose of thinking was to express our gratitude and renew our relationship with creation. These prayers were effective. Our ancient way of thinking held a mysterious power—a power largely unknown to today's rational mind. Whatever we thought became manifest—for thought, speech, and action were one.

We were once the embodiment of what the late physicist David Bohm called "participatory consciousness."[2] We moved with the flow of natural events, coparticipating in their creation. When we danced, our dances were the rhythms of the land; when we sung, our songs were the songs of the earth; and when we spoke, our speech was the animate earth breathing through us. The way we communicated—our ancient languages—were expressions of this original participation.

This sense that humans are embedded in the circle of life—consciously participating in and renewing it through thought—is a very old and once universal idea. It is only in modernity that we in the West began to imagine that our thoughts were separate from the rest of creation.

Is it still possible to recover the ancient ways of interconnected thinking without losing the best of modern thinking? This is what the book explores.

A Radical ReVisioning of Thought

This book is both an explanation of and an exercise in original thinking. This is no small task, because original thinking is a process that requires me to deconstruct paradigms, both my own and that of the greater society in which I was raised. The entire modern Western worldview rests on a peculiar set of assumptions about time, progress, history, education, the role of humanity, and the human economy. In this book, I examine these assumptions in a radical way, but by radical, I do not mean extreme; I mean *getting to the roots* (which is the original meaning of radical) of the way we think.

Without these tacit assumptions, we would see things quite differently. These assumptions govern our lives and are tied to corollary assumptions and beliefs that are similarly buried in our unconscious, or not well thought out. And all these beliefs have consequences. When Western society colonized Indigenous peoples, it attempted to impose a complex set of interrelated assumptions, including abstract beliefs about time (history, progress), land (real estate, nation state), religion (God as a noun), economics (time as money), and so forth. These imposed beliefs all had disruptive effects, not only on others, but on ourselves as well, as we explore. In effect, whatever we do to other peoples we have already done to ourselves.

We have separated ourselves from our lands of origin; we have separated our consciousness from the earth. In so doing, we have gained much knowledge *about* the physical world and developed powerful technologies to manipulate *things*, but lost much of our interconnection with each other, with our own humanity, and with our fellow creatures with whom we share the Earth. Yes, we have developed certain kinds of intellectual capacities to a high degree, but at what cost? When we withdraw into our heads, we withdraw from life. When we see history as a linear progression, we no longer see life as it is—as whole, a circle of reciprocation between human and nonhuman—what we once called the sacred hoop. We have lost touch with our instinctual nature, with our original lands as a source of healing.

A Radical ReVisioning of Time

We have become so out of touch with the timing of nature that we have made time itself into an abstract concept. Time is now something that happens on a clock. And even though clock time often feels at odds with the felt rhythms of our lives, we surrender to the clock as if it were the higher truth. This is one of the most ingrained assumptions of modernity—but it is just a belief.

It is important to realize that our current concept of time as linear is not the way humanity has always thought of time—*and it is also not the only way we can or necessarily always will think of time.* Our beliefs about time influence how we value place, knowledge, and wisdom; they affect the enigma of what it means to be human and the incumbent possibilities and responsibilities of being human, which in turn affects how we educate our future generations and our relationship with the rest of creation. Time and origin are the portals through which the ideas in the book unfold.

Asking the Right Questions

As mentioned in the Preface, the book emerges out of the Language of Spirit dialogues that brought together Western scientists and Indigenous elders. It is because of these dialogues that I was moved to not only offer this book, but also to formulate it in a way that mirrors, as best as possible, the emergence and deepening of thought that occurs in a dialogue circle. A dialogue is built around questions, and so is this book. The purpose of the questions is to deepen thinking. If the intent is successful, then the book will embody the message, and the experience of reading it will be akin to participating in a real dialogue.

The best way I know to get to the bottom of any issue is to ask the right questions—something I learned from Little Bear. Our Western heritage once understood this well, as exemplified by Socrates. The ancient Greeks taught with riddles. I have seen the effectiveness of this method first-hand in dialogue groups moderated by Little Bear. "The only questions worth asking are the ones impossible to answer," said the great Hindu sage Krishnamurti, with humility and wisdom. It is true: the questions that remain alive are the ones that yield the greatest returns. It is for this reason that I have chosen to ask the most challenging questions I could think of, recognizing that the answers can never be complete, but hoping the journey is all the more worthwhile.

The Structure of the Book

The overall story of the book is one of departure from origin—a departure from the timing of the earth—and then a return through the renewal of our original consciousness and true humanity. Thus, the book is composed of four parts: Origin; Departure; Return; and Renewal.

The first three parts, Origin, Departure, and Return, trace the journey of consciousness from our earliest roots until modernity, describing how we came to think in the manner we do today and how we are beginning a return to original consciousness.

Part One, Origin, explores time and the unfolding of consciousness. It reveals how our perception of time changed, leading to a separation from place in general, and from specific places of origin with which we once enjoyed a strong emotional connection. It pries below the surface of what we consider realism and our peculiar notions of linear time and human linear progress, the footprints of modern thinking.

Part Two, Departure, addresses our disconnection from nature from a different point of view, specifically, our notions of humanity and how they have changed. We explore how Western consciousness has come to emphasize our uniqueness from the rest of creation rather than the connections that bind us together with nature. I conclude in a way that shifts the direction away from that imagined uniqueness and toward recognizing our radical interconnection with all our relations (what Dennis Martinez calls our "kincentricity"). This acknowledgment marks the beginning of the return to original thinking.

In Part Three, Return, we delve into the function of thinking itself, which is an absolutely necessary pursuit if we are to discover what kind of world our thoughts have created—and if we ever hope to make deep structural changes to society. I focus on the results created by the myth of separation, considering them anew in light of our journey so far. In particular, I call our attention to the professions that have emerged out of academe, reexamining their origins, and asking what is needed today. I also call us to rethink economics because of the undue influence it currently exerts on all of our social systems and the way it epitomizes abstract separation. Throughout, I seek to reconnect the state of modernity with its life-affirming roots.

The fourth and final part, Renewal, grounds the theory of the book in a praxis I call original education. In this vision of education, the interconnection between subjects is restored. This part of the book represents a lifelong passion for reforming education into a coherent, interconnected whole. Why education? Because to change the world—to make it whole again—we must change the professions; and to change the professions, we must change education because the structure of education determines the professions. But to change education, we must change the way we think. In fact, we must relearn *how* to think, not *what* to think.

How to Read the Book

This book takes up seemingly separate subjects—time, humanity, education, economy, and nature—and weaves them together into an interconnected, but nonlinear, story—a story that is driven by inquiry rather than information. This may be unusual today, but it once was not. And it is done for a specific reason—to exemplify original, relational thinking in the writing itself, instead of speaking about something that is not actually in the book.

The book has coherency and logic, but not logic in the sense of a syllogism; it does not build from one premise to another in order to make an argument. This is why I intentionally do not define original thinking at the outset—not in any detail anyway. Instead, I seek to embody a living process of original thinking without having a definition get in the way. Meaning deepens over time in the book, as it does in a dialogue circle.

The book is thus a journey of discovery. It has its own internal logic—the logic of a tracker. Trackers come to uncover the story of an animal by recreating the totality of an animal's worldview. They do not just follow the flow of direction; they get inside the perspective of the animal, knowing why it moves by how it thinks. In a similar way, I have attempted to track the current state of Western consciousness—how and why we think the way we do—by seeing where our consciousness has come from and where it might be heading. One of the most effective ways to track the origin of consciousness is through the origin of words and their shifting meanings. I examine words and their roots because it is in their evolution that the archaeology of the mind is unearthed.

This is exactly how I suggest you read the book, as a tracker. Read this book with your whole self. Read it not only with your intellect, but with your heart and soul. Read it not as a point-to-point dash down a highway, but as an unhurried yet committed journey along roads less traveled. Read it as a story.

In fact, a story begins each part of the book, to provide a sense of the larger context. Each part then addresses a specific question and is designed to be complete in itself. You can start reading from anyplace, just as you can pick up a track from anyplace. But each part is also a lesser whole, or holon, nested in a greater whole, a larger story. It may, therefore, be more satisfying to read from the beginning because the story completes a circle by book's end.

Each part of the book addresses a question per the following:

- Is it possible to come up with an original thought?
- What does it mean to be human?
- How has our thinking created our world today, and what is now emerging?
- Can education promote the renewal of original thinking?

The first question is intentionally a riddle. It is borrowed courtesy of Little Bear, who posed it as moderator of the inaugural dialogue of Indigenous and Western scientists I was privileged to attend fifteen years ago. All the ideas in the book emerge from this opening question, for it was this question that led me on the journey that culminated in the writing of this book.

PART ONE

Origin: Is It Possible to Come Up with an Original Thought?

The universe begins to look more like a great thought than a great machine.

—Sir James Jeans

Long ago, it seems that there were a Spirit People that became aware that the Earth People were coming. So they decided to have a dialogue … as there was a circle. So all the Spirit People came and talked about this sacred gift, the awareness, where can we put it in the meantime, and then give it to the newcomers when they come. And so they thought about it through the dialogue, and then one said, "Hey, I know where to put this gift where the Earth People won't find it until it's time." So they said, "Let's put it up, way up on top of the mountain, way where they can't climb." And then one said, "No, uh-uh, I don't think so. Those Earth People, oh, they're curious, they're going to find it right away." And they went into circle again. Talk and talk. Finally, one says, "Hey, why don't we put it way under the ground in the cave where the Earth People don't go?" And then they said, "No, no, no. They're going to get it, because that's where they're like ants; they'll climb everywhere." And then finally, they came back together in the dialogue. And then finally one said, "Hey, let's put this awareness in themselves, where they will never find it until it's time."

—Grandfather Leon Secatero

Chapter 1

Original Thought, Time, and the Unfolding of Consciousness

Is It Possible to Come Up with an Original Thought?

I have been thinking about this question ever since Little Bear asked it in the first dialogue SEED hosted in 1999. Like any good dialogue question, one could go about answering it hundreds of ways, and each might bear some fruit, some truth. A funny thing happened when the question was asked, though, which I found particularly significant. Almost all of the Western participants took the meaning of original as a challenge to think of something novel, something that had never been thought of before. The Native American participants, on the other hand, took the question as an invitation to reconnect with a deep place of origin.[1] They went to the heart and root of original, and from that sense, looked for original instructions for how best to live.

It was only sometime after this dialogue, about a year later, when I realized that the very word original is an important key to understanding the divide between Native consciousness and Western consciousness.

To Western consciousness, the meaning of original has become bound up with time and progressive forward movement. Although it can still refer to a place of origin, original has increasingly come to represent something new that has not existed before—which is completely divorced from origin.

The Western view of time has created a distance from place, but to Indigenous consciousness, a separation from origin never even occurred—for if it were to occur, in a very real sense that person ceases to be indigenous, which literally means "connected to the earth," connected to place.[2] A Native understanding of origin is much more of a place than a time.

This realization of the difference in worldviews, as it turned out, was just the beginning of my search. I wanted to know why this split in thinking had come about and if it could be healed. What, in Western thinking, had precipitated the shift in the meaning of origin from place to a beginning of time? And why does Indigenous consciousness, even today, remain connected to origin as a place?

As I pondered this difference, I became more and more intrigued, and many more questions surfaced. Does this mean that Western society is progressing, while Indigenous society is not? What exactly is progress? Could Western society be confusing distance (from origin) for progress? What is the relationship between origin and time and between time and space? What is the origin of consciousness? And what is time? Does time exist independent of any measure of it, or is time an illusory abstraction based on something real (natural rhythms) but not real in and of itself?

By now, it should be obvious that I found the question about original thought extremely generative. All these questions seemed overwhelming and disconnected at first, but as I followed them, common threads emerged. And when I followed those threads far enough, the seemingly separate questions transformed into interconnected aspects of one unified whole. When we dig deep enough into the soil of consciousness, we find a hidden wholeness that underlies different articulations of knowledge. It is this experience of open-minded inquiry—this journey of going down to the roots to rediscover buried treasure—that I would like to share with you here.

How Did the First Awareness Arise?

In 2004, Little Bear asked another kickstart question. This time, he asked: "How did the first awareness arise?" Quickly, however, it changed into:

"How did the original awareness arise?" The word original triggered something in my memory, and I seamlessly traveled back five years to: "Is it possible to come up with an original thought?" It was then that I realized Little Bear's questions were like different facets of a diamond. They appear different on the surface, but underneath they are connected, bringing those of us in the dialogue to the same place from different perspectives. It is a wise and effective way to move the dialogue to a deeper place.

When the dialogue question shifted from a first to an original awareness, something shifted in me. I realized that the distinction made all the difference—or, as the visionary social scientist Gregory Bateson would say, it was "the difference that made the difference."[3] When we say that something is first, it implies a beginning. But saying something is original does not; rather, it implies something that was always there and is ongoing.

Was There Always a Cosmos or Did It Begin at a Point in Time?

Grandfather Leon Secatero once told me in a matter-of-fact way that he knew the cosmos has always been. There was no beginning. This makes sense to me, but most (English-speaking) people in the mainstream believe in a beginning of the cosmos at some point in time, either billions of years ago according to the big bang theory, or a mere six thousand years ago, according to the fundamentalist Judeo-Christian view.

Why is it that we believe in beginnings and, for that matter, endings? It is because of our syntax, according to Moonhawk. "The syntax of English demands beginnings, middles, and ends," he told us in one of the early dialogues, "but this is not the case with Indigenous languages." In Native languages, what in English would be called the past, present, and future tenses can all be spoken of at once. This is nearly impossible to do in English. Perhaps the best expressions we can come up with are the forever now or the ever-present origin.[4]

Do Time and Space Exist Independent of Place?

The idea that time and space exist as independent dimensions is a relatively recent development. For most of humankind's existence, knowledge of time and space was dependent upon place, for it was closely tied to the observation of the natural cycles of celestial and earthly phenomena surrounding one's homeland. Knowing when and how to hunt, gather, and eventually to plant food all depended upon a close monitoring of the recurring rhythms of place. What we now know as time and space were merged into place.[5]

Our modern way of thinking looks back on these so-called prehistoric times as a period defined by a kind of mythical consciousness. The ancients are said to have perceived events as iterations of a cosmic order of eternal return and regeneration within a specific place, whereas today we believe that events occur on an irreversible, linear timeline that is independent of place.[6]

The implication, of course, is that mythical consciousness is just that: *mythical* and therefore unreal, whereas linear time is real. This is an assumption—an a priori belief—that is rarely examined. I challenge this assumption throughout this book, but for now the important point is that ancient Western peoples believed, similarly to Indigenous peoples to this day, that time moves and unfolds in the pattern of a circle.

According to Aristotle, "What is eternal is circular, and what is circular is eternal." The ancient concepts of time were based on celestial cycles, particularly the lunar cycle. This meaning is retained in the use of the word "period" to denote a woman's menstrual cycle. The whole concept of historical periods as distinct from one another is a remarkably recent idea, however. The words themselves do not support this notion of distinct units of time, for period comes from the Greek *periodos* meaning "cycles of time"; and the word "history," which came from the Greek *historia*, originally referred to any knowledge gained by inquiry, not to knowledge confined to a linear slice of time.[7]

The beginnings of what we would now call a "historical consciousness," where events are seen as irreversible, emerged with the advent of

Judeo-Christian theology and its emphasis on an afterlife. But it did not take root for quite some time because the mythical view of time persisted in human consciousness, even in the West. Incredibly, it was not until the eighteenth century that the word history began to be used to track discrete periods of time following one another along a linear timeline. Before then, we did not think of separating distinct periods of history, such as the Renaissance or the Middle Ages.[8] What we call past and future today, which seem to describe opposite poles, once had a more fluid relationship. The past and future once "called to each other, deep calling to deep" and the present *presented* itself like a Greek chorus announcing its arrival.[9]

Our current view of historical time emerged from Isaac Newton's belief in "absolute" time, or time divorced from place. In other words, courtesy of Newton, time became a mathematical abstraction; it is now something that exists independently of space, and something that can be measured in precise intervals. Place, too, has changed; it is no longer something with which we are deeply interconnected. Instead, we derive place by locating space and time on a grid of intersecting vectors of magnitude and direction. This abstracted view became inextricably tied to the West's core perception of reality. It doesn't matter whether or not we are scientists; we hold, as the philosopher Immanuel Kant suggested, a preexisting bias to reorder experiences of the material world in terms of geographic time and space.[10]

Tragically, the Western concept of place has stripped actual places of their once vital energy—an energy that used to nourish us. When we identify a place only as a by-product of comprehending time and space, we lose emotional connection to place. But to Indigenous peoples, a particular place is imbued with collective living energy; it is something that is derived from real events that have occurred there, not something that is derived from blank, soulless coordinates. A place has always been there, and the people and the land have coevolved—which is why, for Indigenous peoples, oral history and geography are inseparable. The stories come from the land and are held within the memory of the land itself.

The inseparability of story and land and the merging of time and space into place were aptly illustrated by the poet Gary Snyder when he told of traveling in a pickup truck through the Australian outback with a Pintupi elder, Jimmy Tjungurrayi. As they were rolling down the road at a good speed, the elder suddenly began to speak very rapidly about the Wallaby people's Dreamtime stories. As one story ended, Jimmy launched into another and then another—and Snyder could not keep up. After about half an hour of this, Snyder finally realized that "these were tales meant to be told while walking, and that I was experiencing a speeded-up version of what might be leisurely told over several days of foot travel."[11]

If a Place Has Always Been There, Is It Primary to Time and Space?

This question seems important within a Western logical context, but it is the wrong question, at least for Indigenous peoples, who view nature as a flux of interconnected elements rather than a succession of first causes. For Tobasonakwut Kinew, the late Anishanaabe elder, a more relevant question to ask is: "What kinds of things want to happen together?" Norm Sheehan, a Wiradjuri Australian aborigine, spoke at the dialogues of places having a "sweet point," like the sweet spot on a wooden cricket bat. The sweet point is where everything comes together: where time, space, and place come together.

Both of these men described something profound, providing a glimpse into the sacred; they speak of how place, time, and space can come together irrespective of which is primary. The best word to describe this in English is "synchronicity," coined by the psychologist Carl Jung to describe meaningful coincidences that do not seem to have a causal relationship. Synchronicity implies a time or place where things come together. It is a curious word in some respects, with interesting roots. It was adapted from kronos, the Greek word for time based on the periodicity of lunar movement; it is also associated with the mythological Kronos, one of the Titans (elder Gods before the Olympians) and the father of Zeus. Kronos is depicted with a sickle, a symbol of the moon.

It may serve us to remember that time was not one-dimensional in ancient Greece, the source of much of Western thought. There was another aspect of time, *Kairos,* which derived from the myth of the divine son of Zeus. A beautiful young man, Kairos is a magical facilitator of the opportune time to act. Once kairos time passes, it can never be regained. Kairos time lives somewhere between the intervals of kronos time—and one needs a fine sense of attunement to recognize an opening into kairos. We get the Latin expression *carpe diem,* "seize the day," from kairos. Kairos is also associated with perpetual motion, as in the eternal motion of the celestial spheres, sometimes depicted as an ever-rotating wheel of fortune.[12]

An Indigenous sense of time, it seems to me, includes both kronos and kairos and then maybe something more. It is understood that all is in flux, that everything is always changing and that even natural rhythms must be closely monitored because they are not guaranteed to remain the same.[13] Monitoring these natural rhythms and cycles helps to develop an intuitive awareness, an awareness that recognizes the opportune time to act within a given cycle. This awareness seamlessly takes into account a host of variables, which are not logical or able to be broken down or counted because they are far too numerous—but they are understood nonetheless at an intuitive level.

Something like this intuitive awareness is necessary for traditional hunting. To hunt successfully, tremendous patience is required—but not merely patience. It is essential for one or more members of a tribe to be in intimate connection with the migration patterns of a particular animal, and these patterns are dependent upon a complex set of relationships involving the animal, other animals, plants, weather patterns, and the like. The connection between the tracking tribal member and the animal is intuitive, composed of far too many variables to be consciously labeled and monitored in a scientific way. Nonetheless, it is possible to know when the animal is near and when the hunt must begin. Similarly, fishermen know where the fish are biting on a particular day, but probably cannot tell you how they know. Anyone can experience this phenomenon, indigenous or not. Rupert Ross, in his classic book *Dancing with a Ghost,* told a story from his own experience wherein he remained for long enough of a time at a particular fishing hole and eventually indigenized himself to that place. He came to know when and where the fish would bite, but he could not tell how he knew.[14]

This sense of timing is especially necessary for Indigenous ceremony, which is conducted when the time is right and cannot be pinned down to an hour or even a calendar day in some cases. In his book *Blackfoot Physics* (which was written after his experiences with dialogue moderated by Little Bear), David Peat relates a story of frustration upon going to his first Sun Dance. Peat explains how he tried unsuccessfully for more than a month to find out the exact date the Sun Dance would begin. He learned only of the approximate times Sun Dances had been held in previous years, and that nothing would occur until after the Women's Society had finished its meetings. Only then could the Horn Society of men meet and the Sun Dance come together. While waiting, Peat witnessed what seemed to be an elaborate set of preparations for the Sun Dance. Later, he chose to understand these preparations as a part of the ceremony itself, and he was able to process the whole experience as a recentering into a wider, deeper sense of time and space.[15]

Clearly, the time of ceremony is dependent on a host of factors that come together in synergistic relationship. This confluence of events is more than likely what Norm Sheehan was referring to as the sweet spot. It is more than the coming together of past, present, and future, and it is probably better to think about it in terms of relationships or alliances rather than time. These relational alliances between human beings and more-than-human beings come together in magical ways, which are outside of linear time or causation as we conventionally think of these.

Such convergences of time and space present an opportunity to restore and renew balance. But what is done in these kinds of situations is rather mysterious to the Western mind. When a synchronicity or a special convergence has occurred and is recognized as such, the Indigenous view does not hold it necessary or desirable to impose a human-centric, egoic explanation. An Indigenous sense of time is not egoic. The knowledge of when and how to act is not about controlling reality; it is about letting a rebalancing occur with human participation, but not control.

It sometimes involves an awareness of when to get out of the way—"to not exist," as Native mystic Joseph Rael used to say in the dialogues—so that larger forces can come through and restore balance. In the process, human consciousness is transported to a larger reality outside the normal Western conception of time and space.

I have been blessed to participate in several multiple-day ceremonies on sacred grounds in which the group has prayed nonstop from start to finish. During these ceremonies, many of us have experienced several mysterious convergences of time and space. I will relate two.

The first occurred after three days of prayer, when we climbed a sacred mountain, made almost entirely of shale. It makes no logical sense to climb a mountain made of shale because the surface is so slippery. That the mountain was also populated with numerous cacti and rattlesnakes made our climbing it all the more logically inadvisable. Yet no one seemed concerned. Those of us who climbed, including several elders in their sixties and seventies, made it to the top, where we conducted a ceremony before descending. Upon our descent, a large boulder suddenly fell from the mountain and headed straight for a few of the descending climbers. Just before it would have struck two of the elders, the boulder suddenly veered, and no one was harmed.

The next morning at sunrise, we were conducting a tobacco ceremony. We had just finished creating a large mound of tobacco when the first sunbeam rose over the hill where the warriors had planted a sacred staff. When the ray of sunlight hit the staff, it was perfectly aligned with our mound of tobacco and the shale mountain behind it, which we had climbed the previous day. At that moment of alignment, a large tuft of tobacco rolled down the tobacco mound in the same arc the boulder had followed the day before, including the same sudden turn. Later, we were told that this was a confirmation from the Holy Ones (the original Spirit People) of our right action the previous day. None of this—neither the decision to climb the mountain nor the decision to make the offering of the tobacco mound—was planned consciously.

The second story occurred the first time I participated in these ceremonies. Not knowing the protocol, I was surprised but happy to be asked to tend the fire near the conclusion of the first evening. I had some idea that this was an important task, but really no clue of how to properly tend a ceremonial fire. I did not know, for instance, that one should feed the fire with conscious intent when adding wood and other sacred medicine, such as tobacco. I also just assumed that someone else would take care of the fire for the duration of the evening. It had already been a long day, and when others went back to their tents, I did too. I soon fell fast asleep.

I awoke in the middle of the night to a voice in my ear. It was Grandfather Leon. He was not physically there, mind you, but it was unquestionably his voice, with its inimitable soft and reassuring tone. He said, "Glenn, get up now and tend the fire." And so I got up, put on my shoes, climbed out of my tent, and proceeded to almost fall over—literally floored by the immense canopy of stars. Coyotes howled—at first from miles away—and then from right over my shoulder. I nearly jumped out of my skin. Damn! They must be just outside the camp.

I gathered myself and refocused on the task at hand. I could make out the fire in the distance, and carefully picked my way through the terrain toward it. I did not use a flashlight, as I find I see better without one, my eyes quickly adjusting to the ample starlight and the fire. When I arrived, the fire was indeed dying, desperately in need of food. I carefully loaded the wood onto the fire until it returned to a full blaze. And then, not knowing any better, I went back to sleep, trusting that someone else would watch from then on.

In the morning, the fire was still going, but Grandfather Leon was not there. In fact, he was not there during the night—not to sleep next to the fire and tend it, as he normally did, nor to whisper in my ear. I was told he had gone back to town—a trip of several hours one way, because he had gotten a message. It could not have been a cell phone or text message because those things did not work where we were. It must have been a psychic message, one that told him that someone in our party had become lost and needed his help. And indeed, he came back with the wayward visitors a short time after dawn.

I couldn't help myself from musing, "How could he have known to tell me to tend the fire?" He was nowhere near the camp when I heard his voice. So how did he do it? I tentatively concluded that he must have bilocated his consciousness, observed the dimming fire, and then spoke through me. Later that day, we were out for a hike, and I took the opportunity to ask Grandfather Leon about it, but he declined to answer. He simply pointed to the contours of the canyon we were then approaching, which, in itself, was a message, telling me to let go of rational thought and just experience. Leon's son, Orlando, trained in the old medicine ways, eventually offered a different possibility. He suggested that it was the fire that spoke to me directly, using Leon's voice because it knew that would get my attention. I found that possibility the most intriguing.

The next day, sometime after breakfast and morning prayers, I again felt an unusual presence. I looked up in the sky and saw two immense figures, one male and one female, stretching across the sky about fifty times larger than the sun. I was not under the influence of any drug other than some gritty morning coffee brewed over the fire, but I saw these figures as clear as day. I believe they were the Holy People themselves, revealing themselves to me for a moment, offering a special blessing.

These kinds of synchronicities and blessed protections occur, seemingly most often, when egoic, controlling thought gives way, surrendering to a larger power. This is a form of original thinking that emerges in participation with a realm of natural forces beyond human will. It defies explanation, but it happens.

What Happened in the West to Change Our View of Time and Space?

In ancient Western philosophies, there was an awareness of the need to surrender one's will to nature's timing. Although a vestige of this awareness is retained in certain religious teachings, something happened in the West that changed the predominant way time and space were conceived, and this change was the precursor of further changes in consciousness, which profoundly affected the way we lived. Specifically, it changed the way we thought of ourselves as human beings and our purpose on Earth; it brought about the development of ego, abstract and rational thought, perspective, and so forth. It is important to realize that these latter changes were not developed by Western culture alone. All human beings developed these modes of consciousness. But, in the West, we chose to emphasize these emergent modes of consciousness as the pinnacle of progress, and that is what cemented our path.

To adopt this new emphasis, we had to change our view of thought as coming from nature (as in coming from the fire in my story) and instead embrace the idea that we humans are the originators of our thoughts. We had to separate consciousness from nature and locate it in the human brain. We took the wide window of consciousness we once located in the entire natural world and narrowed it into our individual heads. In doing so, we eventually came to believe that we are the only conscious creatures on Earth.

We retained a vestige of appreciation for the spirit of nature, but this largely moved into the province of what we could no longer explain and was labeled "supernatural." Indigenous peoples, on the other hand, chose not to emphasize these abstract modes of thought, and so the supernatural remained *natural*. Indigenous peoples have never accepted the idea that human beings are separate from nature. As a result, they retain a wider window of consciousness as their norm.[16]

What Is the Benefit of Abstract Thought?

Abstract thought is a mode of thinking that enables us to imagine particular scenarios in our mind's eye and then to design a physical vehicle for what we have imagined. Aristotle referred to it as the human ability of "deliberative imagination" (as opposed to "sensitive imagination" which he thought all animals have).[17] It is a powerful tool available to all human beings, but one that has been particularly emphasized in Western consciousness.

Abstract thinking freed the West to develop a formal way of knowing—modern science—which studies particular aspects of reality in order to reliably predict and control much of our experience in any place or time in the future. At its core, science is an exploration of the unknown.

These explorations result in spin-off technologies, which are an application of the known[18] that can potentially improve the human condition.

Does Abstract Thinking Have a Shadow Side?

Abstract thought is by definition an abstraction, or a partial impression that is extracted from the whole of the landscape. Knowledge obtained in this manner is no longer connected to the cycles, natural rhythms, and pulses arising from a particular place. When knowledge is no longer rooted in place, it is a short step to believing that knowledge can be transported to any place. This is the seed of imperialistic thinking and colonization. It led us to see progress as including the idea of expanding our reach into foreign territory (manifest destiny).

For the past five hundred years, abstract thinking has been a force for both creative initiative and its shadow, imperialism. Most of us would regard the emphasis on undiscovered frontiers of thought to be a good thing. But is it always? What happens when thinking is one-directional, always moving forward and not rooted in tradition? Does this create an imbalance? I believe it does. The philosopher Kierkegaard spoke of living our life forward, but understanding it backward. In other words, we need the totality of past, present, and future to make meaning of our lives.

Is There Anything New under the Sun?

Thinking that is overly forward, constantly moving ahead like a shark without looking back, is dangerous. It does not allow us to learn from the past—neither from our accomplishments nor from our mistakes. We delude ourselves into thinking "this time it will be different." Many economists do this, for example, when they ignore the signs of an economic bubble. Is it not safer to assume that economic phenomena are cyclical, and thus to prepare for a downturn following a rise and an upturn following a fall? "In peace, prepare for war; in war, prepare for peace," said Sun Tzu in an ancient text.[19] "What has been will be again; what has been done will be done again. There is nothing new under the sun," is a biblical teaching. Western consciousness ignores its own wisdom teachings, along with the perennial wisdom of the globe that stresses balance. The word "perennial" is itself an expression of inevitable return, as in a perennial flower, which returns to grow again in its season.

All of nature occurs in cycles. Why wouldn't the actions of human beings also occur in cycles?—for we are a part of nature. We work against our nature when we move only in a forward direction—keeping up with the latest and greatest technological innovations at the expense of perennial wisdom. When the emphasis is always on the new, it is easy to fall into piecemeal or reactive thinking; the focus becomes limited to what has most recently occurred, without considering the full context of history. The educational reformer John Dewey, upon reflection, admitted that this happened with progressive education. Progressive educators tended to react to aspects in education that they had identified as deficient. They would improve upon those areas, but neglect other foundational elements of education, because those elements were essentially being taken for granted and were no longer seen. Accordingly, the implemented changes created new, unanticipated problems.[20]

This kind of thinking—attacking symptoms without identifying and addressing the underlying systemic issues—occurs all the time in modern business and in many other arenas, and has been termed "shifting the burden" by Peter Senge and other organizational consultants.[21]

Chapter 2

Looking Backward to Go Forward

The European Renaissance is often considered the seed of the modern scientific era. Although there is some truth to this, what is less understood is that the Renaissance began as a movement that sought to look backward in order to go forward with a more complete and whole vision. The symbol of that movement was the Roman god Janus, the god of doorways, who looks both ways and whose name is the root of "January," the first month of the calendar year when we look back at the year we have just completed and also ahead to the new year.

The Renaissance, and the scientific revolution that unfolded from it, brought us some of the West's most creative and comprehensive thinkers, including Leonardo da Vinci, Galileo Galilei, and Sir Isaac Newton. Although these men were the precursors of modern science in certain ways, they are often miscast by historians as the first scientists, in a thinly veiled attempt to justify the dominance of science today. In truth, they were known in their time as natural philosophers—people who loved wisdom and nature. The term "natural philosophy" was used from antiquity up until the nineteenth century, when the word "scientist" was first coined.[1] The word science was used long before, but it referred to a general pursuit of knowledge, coming from the Latin *scientia,* which means knowledge or to know. There are many ways of knowing, and modern Western science is just one of them.

A natural philosopher was a thinker concerned with the whole field of nature. One of the reasons da Vinci could be da Vinci was that, in his day, there was not a separation between what we now call the arts and the sciences. Beauty, harmony, and proportion were (and, in truth, still are) integral to both.

For the natural philosopher, the pursuit of knowledge was not a utilitarian pursuit so much as an aesthetic one. Natural philosophers looked closely at nature to observe its patterns, which were continually unfolding, dissipating, and then recurring. In so doing, they joined with people all around the globe and throughout the ages who have embarked on such study and observation. But what made the European Renaissance such an exciting time was the way in which these understandings—many of them ancient—were incorporated into contemporary art, architecture, and mathematics. The beginning of the Renaissance heralded a healthy balance between old and new, at least for a time.

Yet something happened in da Vinci's lifetime that laid the framework for what would become a hallmark of Western consciousness: a fixation on the future. What occurred was actually the culmination of what had been creeping into human consciousness for nearly two millennia, fed in part by religious ideas of a better future awaiting us after death. This gradually unfolding consciousness was encapsulated in a single breakthrough, and ever since, Western consciousness changed in a way that makes it difficult to remember how it was before. That critical breakthrough was what is known in art as "perspective."

What Is It about Perspective in Art that Emphasizes the Future?

Perspective, as any art student knows, is the technique of drawing a landscape from the single point of the eye in sight lines that recede away from its vantage point. Things closer to the viewer are larger and therefore assume greater importance, while things farther away are diminished. The development of perspective freezes the point in space and time from which is being viewed.

Although in reality we have two moving eyes, the perspectival view of a scene is derived from a single stationary eye. The lines of perspective literally represent the future, which is off in the distance; in other words, the future is facing forward. This development marked a dramatic shift in human consciousness. It may not appear to be a revolutionary development, but it was, because it reified a shift in consciousness that had already begun to occur.[2]

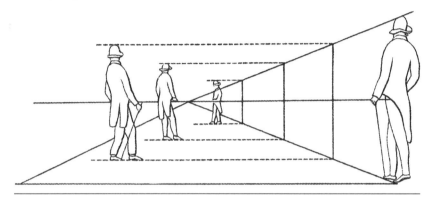

Figure 1: Linear perspective

The ancient Greeks, for instance, believed that *the past was ahead of them*. For them, this was only natural because they could see the past actions that had already occurred. The *future was behind them* because they could not see the future, which moved up and enveloped one from behind.[3]

That shift in perspective changed the orientation—and that changed everything. Once humans became convinced that we could see the future in front of us, we began to think we could bend it to our will. We shifted from *original participation with* nature to *seeking control over* nature. We started to imagine that we could be the masters of our own destiny. The ancient Greeks called this *hubris,* or the tendency of humanity to elevate itself above the gods. Today, however, it is just called science, or the ability to make calculated predictions about the future based on data obtained from the past.

Once the West adopted the emergence of perspective as the most realistic view, many other changes in the West's worldview were set in motion. The philosopher Jean Gebser, writing in the mid-twentieth century, equated the modern worldview with this "post-perspectival consciousness."[4] Before perspective, it was believed that energy came from nature and that soul resided in the world, in what was called the *anima mundi,* or world soul. After perspective, the human eye and consciousness came to be thought of as separate from all else; humans became the detached observers of what eventually came to be seen as a disenchanted and desouled world.

Before perspective, human beings were an integral part of the world, not set apart from the world. After perspective, human beings become observers of their environment. Before perspective, spirit and thought were understood to come from a place outside our individual consciousness; we could be inspired, and we understood that the source of our inspiration was beyond us. After perspective, we came to believe that we alone are the generators of our thoughts. We became individual souls separate from nature, one another, and all other life forms. And, in the process of creating perspective, we unbridled an untrained horse within us, a force that came to be known as the ego in the eighteenth century.

The concept called ego, which we locate in our intellect (or head), reinforced the shift that occurred with perspective. We began to see the world from the eyes in our head, like the proverbial "Fool on the Hill" in the Beatles song; we dismissed or downplayed the sensations through and within our bodies. And we began to imagine that we humans are the only conscious beings and that all the rest of creation is unconscious, with the exception of other humans and animals. But even here, we are stingy, attributing an inferior consciousness, sometimes begrudgingly, to animals and even sometimes to other humans.

This worldview is so embedded in the consciousness that it lives below the surface of our attention, making it extremely difficult to discover. We do not easily realize that the aggrandizement of ego has shifted the credit from nature to ourselves. The world is now deemed dead unless or until we direct our consciousness upon it, as if our gaze is what brings it to life. But even then, when we are engaged in the observation of nature, we do not realize how we are, in effect, stopping nature in order to take a measurement or a snapshot of it. We stop the unfolding rhythms of nature and then imagine that our projection of consciousness unto nature is the only motion occurring. We create time as a line, and claim it as the sole province of human consciousness. *Time and ego become one.* In this way, we have taken time away from nature, away from the Gods, and claimed it for our own.

I am explaining this shift in consciousness carefully because we normally take our worldview for granted, unable to see that our worldview is truly just a view—a certain perspective that systematically incorporates certain areas of perception while diminishing or erasing other ways of seeing. The Western world followed a path that supported innovation through the mastery of nature, which unquestionably led to remarkable achievements that, in turn, reinforced confidence in this path. As we continued to create greater and greater technologies, it became nearly impossible to see the validity of any other culture that believed and acted differently.

Since time became a forward line, we have seen all our ingenuity and inventions as *progress* along that line, making it easy to jump to the next level of belief: that we are at the head of the parade. Any other worldview that does not agree with ours must then be behind us! The truth, however, is that all worldviews are partial. Some have a wider window of consciousness, but all have blind spots.

Is Not Viewing the Future as behind Us Still a Linear View?

Yes, to some degree. But the ancient Greeks actually thought of the future as moving up on us, not only in a line, but from all directions. According to Moonhawk: "While we moderns navigate a rather static space-time worldview by moving through time, *time moved through the Greeks:* circling the world it came up to them from behind, where we don't have eyes and can't see things coming at us, moved through them and into the past in front of them, where they had eyes and could see" (italics added).[5]

The key difference is that, for the Greeks, the future was unseen and still mysterious, because it was behind them, and therefore they did not imagine they could control it. But we moderns believe the future is ahead of us and have no compunction believing that we can control our destiny, that our fates are limited only by what we imagine we can accomplish. The shift that came with perspective and a focus on the future is profound. It marked the beginning of an addiction to progress that became overpowering with the industrial revolution.

The Industrial Revolution

Some two hundred years after the discoveries of Galileo, Newton, and Descartes, spin-off technologies emerged that facilitated the rise of the industrial revolution and the modern era of nation-states and global capitalism. The idea of looking backward to move forward receded in memory with the explosion of powerful new technologies and the profits they brought. Government and industry had a vested interest in selling people the promise of progress. Not surprisingly, it worked. People soldiered on, working to improve the standard of living, not only because they were captivated by the hope of progress, but also because they were resigned to the inevitability of progress. In other words, they took the bad with the good.

The distribution of goods and services increased greatly during this period, even as many people put up with abhorrent working and living conditions. Such conformity could not have occurred without the prevailing notion of inevitable progress, a belief and value that remain deeply embedded in the modern Western psyche. Because of this strong belief, we committed ourselves to the full-fledged exploitation of nature for the good of the economy. Only now has a seed of doubt taken hold to challenge this notion of unimpeded growth, because we can finally see that it is ecologically unsustainable. But at this point, the entire ecology of the planet is at risk. With many of the so-called developing nations wanting to follow the course of progress as laid out by the West, it is difficult to imagine how to reverse this inertia that still continues from the industrial revolution.

To change course, we will have to pause long enough to disengage from our fixation on the future and reevaluate our purpose as human beings on this Earth. And, because our view of time lies at the very root of the current condition, the greatest hope may lie in rethinking how we think of time.

Hasn't the View of Time Changed with the Theory of Relativity?

Yes and no. With the advent of the general theory of relativity, time and space became inseparable, and time was no longer seen as absolute. In other words, the idea that there are precise intervals of time that can be measured between events, from any point in space, was overturned. Einstein demonstrated that time is relative to the observer, and in doing so, he also demonstrated that the universe is not static. Space and time are dynamic quantities, and bodies moving in space affect the curvature of space and time—and vice versa. This implies that the universe is an undivided whole, which is a break from Newtonian physics, but consistent with what the ancients believed.

Nonetheless, it is important to bear in mind that special and general relativity theories retain much of Newtonian physics in their formulations. Space and time are seen as dynamic quantities within a field, for instance, but the field is still seen as having a fixed position in space-time. Thus, relativity theory retains the implication of beginnings and endings, and it does not change the Western worldview of time as progressive and moving forward.

With the Hubble telescope discoveries, the universe is seen as dynamic and expanding from a single point or big bang singularity. A point of beginning implies an ending. Because the universe of the present is believed to be a result of billions of years of expansion, big bang theorists postulate that the origin of the universe must have been a quantum singularity involving infinitesimally small compressed space. This is where quantum theory—the other great scientific theory of the twentieth century—is used to explain the subatomic realm of the universe's mysterious origins.

It should be noted that Einstein never accepted many of the principles of quantum theory. As Little Bear likes to say about "Grandfather

Einstein," "He came right to the door, but didn't walk through the door [to embrace quantum theory], because he didn't want to be known as the mad scientist."

But for Native Americans who have been exposed to quantum theory, the opening of that door was like coming home. It made Little Bear realize that Western science had, at last, come full circle to something resembling a Native worldview, and it provided the foundation for the dialogues between Native and Western scientists to begin. The advent of quantum theory begins to truly break down the Western concept of linear time.

How Does Quantum Theory Break Down the Concept of Linear Time?

When we looked inside an atom and first viewed what has become known as "the quantum realm," we realized that we could only know an atom by its radiating, but not by any distinct properties we could stop and measure.[6] The three major features of quantum theory—discontinuous movement (known in popular language as the quantum leap), the particle/wave paradox (now it's a particle; now it's a wave), and nonlocal entanglement (separated particles adopting a complementary spin in relation to each other no matter how far apart they are)—are all predicated upon a worldview that does not see phenomena as separate and distinct entities in space-time. In other words, there are no longer any *things* in the quantum realm—just process and relationship.[7] Think about that for a minute!

Unfortunately, wrapping our minds around this notion is what makes quantum theory inscrutable to many people. But in my experience, this is not difficult for Indigenous people to understand because they have believed all along that every aspect of the cosmos is radically interconnected, without time. Maryboy spoke about this in the dialogues, noting that when a prayer is sent to all the directions in Native ceremony, it is understood to travel everywhere instantaneously. It is possible to connect with origin, or the source of our creativity, at any time because it is truly outside linear time.

Quantum theory is, in many ways, consonant with a traditional Indigenous understanding of the cosmos. It is literally a revolutionary idea in that it has completed a full circle of revolution back to its origins. Because of this, it is also completely at odds with everything held near and dear to the present (and dying) paradigm: that time is a line, that solid objects exist alongside other solid objects in space, and that our consciousness is separate from the things we observe. Quantum theory overturns all of this, which is also why it has been slow to penetrate into everyday awareness.

Nonetheless, quantum theory, like any scientific paradigm, did not develop in a vacuum. Many in our culture seem to be experiencing a rising wave of consciousness, and its purpose may be to disentangle us from our limited view of time and space. I believe a new paradigm is starting to unfold, which of necessity requires us to break down much of the old. It is especially challenging our fixation on linear time. This may be why we have been fascinated in this past century with things that altered a linear perspective or a linear beat—from the artwork of Matisse, Picasso, Braque, to ragtime, jazz, blues, and rap, and even to ideas such as time travel. This may also be the source of a renewed interest in Indigenous ways of knowing.

The paradigm shift may not yet be in place, but it is coming—and there is nothing that can be done to stop it. It is, ironically, only a matter of time. And, as consciousness evolves, so does our perception of time because the two are inseparably linked.

How Does Consciousness Evolve?

Jean Gebser, the philosopher Ken Wilber, and others believe that humankind has evolved through specific stages of consciousness—archaic, magical, mythical, and mental—and that we are now giving birth to an integral consciousness or, as Sri Aurobindo termed it, a "supramental" consciousness. I sense this is true. I also believe it is important that we do not imagine these new modes of consciousness to be eclipsing or supplanting previous structures of consciousness because that thinking will trap us once again in a linear view. Human consciousness is best seen as an unfolding of potentiality that very much includes and retains the former in the unfolding of the new. Aurobindo emphasized this in his recapitulation of ancient Vedic knowledge. Gebser also understood this, even though the majority of Western philosophers believed otherwise, emphasizing the emergence of new structures of consciousness as if the old became obsolete with the arrival of the new.

The new does not supplant the old; it only reorders our awareness. This restructuring can be more or less fluid depending upon our capacity to integrate the whole. This is why psychologists refer to the living energy of archetypes that can guide us in the present day. In any given era and in any given culture, one aspect of consciousness tends to be emphasized while other aspects are suppressed or minimized. Particularly in Western cultures, we label the emphasis progress, but it is not really progress— only distance from the origin. The concept of linear progress is largely an illusion.

Chapter 3

Wheels within Wheels

What Is Progress?

All we see as progress is, in actuality, an unfolding of inherent potentiality. The idea that we progress along a line of development away from our origin is literally misleading. Just as a seed holds the potential to become root, bud, and fruit, the origin remains the source of our creativity even if we imagine otherwise. I am sharing a teaching that has come down through the ages from many peoples and places. We have a divine essence that unfolds from the core of our being. Jesus called it the kingdom of God within; Buddhists call it buddha nature, and the ancient teachers (rishis) of India called it atman, or the higher self.

This ancient teaching from India gives perhaps the most lucid and comprehensive explanation of the course of consciousness. A day in the life of Brahma is said to be a grand cycle of consciousness or Manvantara that originates in pure spirit or *purusha*. Purusha then enfolds itself (involution) into prakriti (matter) by descending through structures of consciousness called Sat (Being), chit (consciousness-force), ananda (bliss), and through the intuitive gnosis, mental, emotional, and gross physical stages.

As purusha descends, it travels through increasingly dense forms of consciousness until at the bottom of the cycle it is encased in matter (prakriti).[1] It then returns, unfolding (evolution) in the inverse order in which it enfolded.

Variations of this great circle of life are found in virtually all wisdom traditions; it is an ancient understanding that could potentially unify humankind just as it unifies matter and spirit. This notion of spirit interpenetrating matter, and matter returning to spirit, for instance, is related to the Indigenous idea that one can become the other—and back again. It is at the root of why Indigenous people refer to rocks as "Stone People" or "Rock people." According to my Native American friends, rocks have spirit. They also have thought and memory, they say, which is why they are sometimes referred to as "Grandfather rocks." This is a stretch for Western people to imagine, but why not? In Vedic cosmology, physical matter retains the knowledge of the entire inward arc (involution) that precedes evolution, so they are our grandfathers in a very real sense (see Figure 2).

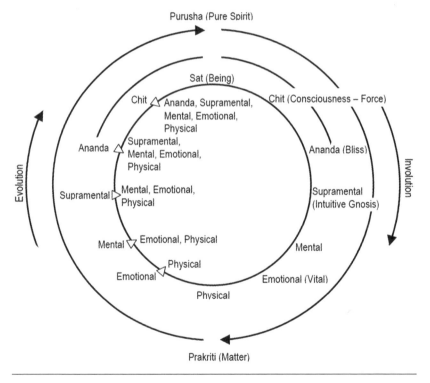

Figure 2: Involution and Evolution

In Vedic thinking, when evolution begins, it is actually the unfolding of Spirit that was always present, but was enfolded in consciousness until its proper time to emerge. Spirit and matter are thus intertwined and codependent with each other. The Hindus have a saying that "Purusha is like a lame man with functioning eyes carried on the shoulders of Prakriti, a blind man with functioning legs."[2] Only when they are together can the walk of evolution proceed.

The evolution of humanity is thus a part of the overall creative urge of Being. It is not a linear progression in which structures of consciousness emerge apart from the rest of creation. No, I believe it is all interconnected, and that the myth of linear human progress is just that—a myth—because everything that evolves has been previously involved (enfolded) in consciousness. Everything that comes into being, in this view, is simply an unfolding of what is already there lying dormant until its proper time to manifest.

This understanding is embedded in our perennial wisdom. The lotus plant is considered sacred in India because the unfolding of its beautiful petals is likened to the unfolding of the *atman,* or the beautiful essence of our highest self, which we carry within us as latent potential until it is time to awaken to our true nature. Hidden within human development and evolution on this physical plane is the impetus of spirit. This is why we experience the yearning for spirit as a coming home, or a coming full circle. As the Russian seer Madame Blavatsky understood, the human being is a "spark from God" [origin] on its return trip thereto.[3] So, too, many cultures did not see progress as a line or a single circle, but as the unfolding of circles within circles, or a spiral progression. The end of a cycle leads to a new cycle, and the joining of two (or more) cycles creates the spiral, which has been a sacred symbol among all cultures for as long as we can remember. This cross-cultural symbol originally came from nature. Nature retains the spiral in many magnificent examples of its unfolding processes (of fractal progression), such as the seashell, pinecone, leaf, flower, vegetable, and seed head.

Figure 3: Fractals in nature

Does the Perception of Space and Time Change as Consciousness Evolves?

Within the framework of the Indian model of involution and evolution, the perception of space and time is continually changing. As Spirit begins its descent into matter, space predominates, and time is relatively quiet (feels more eternal). However, as we approach the midpoint and beyond, time gains prominence over space. Space becomes the background, and time seems more real, more pressing.

In modernity, we are past the midpoint on the Vedic cycle—which is likely why many of us today think time is speeding up. Those of us who are fixated on speed are generally more agitated in our thinking, because we are enmeshed in material reality. But it is possible, even today, to achieve a state of deep calm. And when we do this, we are closer to the spirit of nature, and we experience a sense of spaciousness more real than time.

I learned about this as a child, from, of all things, a bee—an animal we normally associate with busyness. When I was all of about six years old, I got stung by a wasp, and my arm swelled up. My mother told me I was allergic to bees, and to avoid them. I soon became afraid of bees, but not for long. Shortly thereafter, I had a dream. And in my dream, a bee came up to me and gently whispered in my ear. He said, "Be(e) calm." And ever since, I have been calm around bees, and more calm, in general. But I still have much social conditioning to overcome around being productive.

I have been describing both a general shift in consciousness that is connected to a universal unfolding of nature and the perceptual choice involved. The perception of time and space, however, is never simply an individual choice. It is culturally determined, or at least mitigated by cultural choices.

Asian and Indigenous cultures have, relative to the West, continued to emphasize the world of phenomena as transient or impermanent. They have emphasized this in different ways.

Indigenous cultures have maintained a greater level of participatory consciousness in relationship with the ever-changing flux of nature. Asian cultures, once engaged in the same level of participatory consciousness, but then began to distrust the material world, seeing it as *maya,* or illusion.

Western cultures have gone through similar stages of original participation and then distrust, but somehow emerged with near-total conviction that the material world is the most important reality. The West has come to accept objective reality as the truth rather than a mere representation of the higher truth, as Plato taught. The Western world-view is most involved with phenomena (with things). This has yielded great knowledge of particularized reality at the expense of losing our living connection to the whole.

How Is the Western Perception of the Material World Changing?

The worldview that has permeated modern Western consciousness ever since the advent of perspective in the Renaissance is now breaking up. With it, the perception of linear fixedness that kept us focused on discrete material reality is being torn apart. This change can be experienced as liberating, but more often than not, it is confusing. It can be hard to understand why everything is breaking down or to know what is coming next.

The breakup of a worldview is like having the ground shift under our feet. We are witnessing cultural discontinuity and fragmentation in multiple forms, within art, architecture, music, film, literature, and even within governments, as evidenced most recently in the Middle East, with what became known as the Arab Spring.

Without an understanding of what is happening, all these changes are lumped together and labeled the "postmodern movement"—a catch-all moniker to describe a reaction to the confining worldview of modernism. The postmodern movement is not really a movement as such because it is largely disjointed and incoherent. It is an interim state emerging out of long repressed emotionality that, in my estimation, conceals a hidden desire to return to a state of wholeness.

The only thing certain in these times is uncertainty. Since the mid-twentieth century, we have been in a tumultuous period that has created enormous stress and change. The advent of the atomic bomb threatens humanity's very existence and that of many, if not all, species on Earth. The same can be said of the human-made pollution of our water, air, and land. The global climate is changing rapidly, already endangering the habitats of human and animal species near the poles and at high elevations.

The rapid pace of such changes is felt as a quickening that some interpret as the end times. But it is mostly the addiction to linear time that makes it feel as though this is the end.

In my view, it is not the end, only the end of a cycle. We need to go through the chaos, so that we might question the path that it has been on and yield to a new (an original) consciousness. The perception of time and space is changing, and this change literally represents (re-presents) something original, which is both new and a return to an ancient way of seeing.

It is not necessary to completely reject the previous paradigm to birth a new one, nor is it necessary to reject material reality as unreal, as some cultural and religious traditions have done. What is necessary is to pause long enough to question our preoccupation with the material plane. The suspension of our habitual way of seeing is what makes it possible to awaken to the hidden wholeness underlying the fragmented world of manifest reality.

Bohm called this unmanifest world of potentiality "the implicate order" and the manifest world "the explicate order." The implicate order precedes the material world; for material reality to be born, it must first be conceived in spirit. The implicate order could just as easily be called spirit, God, or consciousness. The name is unimportant. What is important is the understanding of a subtle realm that pervades nature and underlies all there is or will be.

It is no coincidence that Western peoples began to explore wholistic[4] consciousness around the time the atom was first split. Splitting the atom was symptomatic of our deeply split and fragmented psyche, and we needed to be whole again; we needed healing. When Western peoples began to question the material realm, it led to a vast exploration of perennial wisdom all over the globe. We turned first to Eastern wisdom and eventually to the Indigenous wisdom that was here in the West all along. What has happened in the Americas is only a part of the emergence of a worldwide awakening of intuitive awareness, a harbinger of things to come, as the predominant rational mind begins to lose its suffocating grip.

The limited thinking of rational mind has caused great suffering because rational (perspectival) mind objectifies the world and distances us from it. This creates a self-imposed tear from the fabric of the universe, from the source of our collective healing—our original mind. Over time, we have played a mental trick on ourselves. We have come to believe that the abstract representation of reality is reality itself. But it is not reality— or at least not the whole of reality.

Those of us who cling to a materialist worldview only feel secure when we stop and measure reality. Tellingly, the root of the word "measure" is from the Sanskrit *maya,* meaning illusion. But we cannot contain and hold the whole of nature in our material brains. Nature is too big. While she presents one visible face that we know as material reality, she conceals enormous levels upon levels beneath our sight.

We once accepted that we could not know the whole, and embraced all of life as a Great Mystery. But when we became accomplished in our manipulation of the material world, we forgot the underlying spirit of things, the original source from which all came into being. Now we suffer because we sense something missing, but we do not know what it is. We compensate for our dissatisfaction and emptiness with material pleasures—and though this is an acute problem today, it existed long ago as well. In Buddha's time, he taught that the way out of suffering was nonattachment. Today, however, we are conditioned to be attached; we are taught to pursue happiness as if it is something outside ourselves, something that can be acquired in the marketplace (an illusion that is fueled by ubiquitous advertising). But we will never find happiness through its pursuit; we will only find happiness when we rediscover what is already whole and complete inside ourselves. Inside we possess original mind, the ground of being we all share.[5]

Indigenous teachings that recognize the cosmos as already complete and whole help to break down this addiction to material cravings, as do Eastern meditation practices that teach us to pay attention to the thoughts we habitually cling to. Letting go of excessive thinking and accepting *what is* moves us, however slowly, toward inner peace. We begin to realize that mental activity alone will not solve our problems—that, in fact, it tends to create more problems than it solves. On some level, people are increasingly sensing that there is something deeper underlying our superficial thinking.

Original thought emerges from a state where thought and nature are one. Long ago, all thought was considered sacred. People used to be, as Indigenous elders are still today, extraordinarily careful about what they thought and what they said. We are relearning that it is important to monitor our thoughts and to return to that original state in which thought and action are consciously joined. A first step to rediscovering the sacredness of our thoughts is to allow our thinking processes to slow down. Slowing down allows a deeper knowing to come through. This naturally happens in meditation or in a well-run dialogue circle. We can all practice slowing down, with or without a teacher.

Ultimately, our thoughts are not "our thoughts" because thoughts belong to nature. Thoughts emerge from nature and, depending on our vibrational resonance, we attract certain thoughts to us. But we tend to think of thoughts differently; we believe we own them and generate them in a wholly intrapsychic way. In my experience, this simply is not true. As I related before, I have participated in dialogue circles where it is not an infrequent occurrence for the very thoughts I am thinking to come out of another's mouth. I have learned that my thoughts are only my thoughts to the extent that I identify with them and allow them to hang around me.

Our thoughts are not really ours to own. Thoughts have a life of their own. If we are attracting destructive or nonbeneficial thoughts, we can and should release them. Like the air, water, and clouds, the nature of thought is to move. When we release *all* thoughts (beneficial or not) and clear our minds, we prepare ourselves to attract original thoughts that emerge from the heart of nature. We move closer to our original state of oneness, the state of original mind. We become one with the flow of nature and experience peace.

Where Are We Now in the Cycle of Consciousness?

At the present time, we are in what Gebser called the deficient stages of the mental, or rational, structure of consciousness. We developed both abstract and rational thought processes, resulting in myriad innovations, both dramatic and subtle, that have improved our economic standard of living and served us well for a time. But these capacities have outlived their usefulness.

We now spend an inordinate amount of time cogitating away in our minds without being present to the natural rhythms around us. The ingenious ways we have devised to survive without awareness of these natural rhythms have also caused us to lose something of our innate humanness. We have become lost in thought, confused and disoriented. These are difficult and turbulent times for us physically, emotionally, and spiritually. There is disharmony throughout the globe as cultures that think and act differently encounter one another.

The good news, however, is that we are approaching the end of the era of rational mind as the predominant mode of consciousness. The beginning of the unfolding of an intuitive (and more feminine) way of knowing is upon us. Rational thought, frequently associated with the masculine principle, will not go away, but it will no longer be our master. The emerging integral consciousness will include our physical, emotional, and mental structures, and these will underlie a new more inclusive understanding (wherein the rational will be literally *standing under* the intuitive).

These are confusing times because they are transitional, but they also offer an opportunity to let go of old patterns that no longer serve their purpose and to open to the larger spectrum of what is happening. According to Grandfather Leon, this is the time—"the next 500 year calendar"— when we need to put aside the negativity and confusion of the past and remember our connections with all human beings and all creatures. A new day is upon us, and according to Grandfather Leon, the only way we can make the transition successfully is to wipe away our tears and walk hand in hand together, with one mind and one spirit, as the "five-fingered ones," as human beings. We need to recapture what it means to be fully human in order to usher in a new era of integral consciousness, in which the full spectrum of our human potential is activated.

The Mayan Calendar

In 2012, all we heard about was the Mayan calendar. For me, this was especially true because I helped organize a conference on the subject that was called "Wisdom from the Origins: The Mayan Calendar and Other Prophecies on the Future of Humanity." The conference committee invited many noteworthy Mayan elders, such as Don Alejandro Cerilo Perez Oxlaj (the Head of the Mayan council of elders of Guatemala who was introduced to me by Grandfather Leon years ago); Grandmother Flordemayo (a Mayan priestess who studied with Don Alejandro and is now one of the members of the International Council of 13 Indigenous Grandmothers); Don Gaspar (in the lineage of genuine Mayan kings); Tata Pedro Cruz; and an impressive young Mayan seer, Ac Tah, among others.

As it turned out, none of the Mayans believed that December 21, 2012, was going to be the end of the world or even necessarily cataclysmic, but all of them believed that this time was about major change, and that it offered a special opportunity to learn to be comfortable with change.

By the end of 2012, I grew a tad weary of end-of-the-world jokes. People kept telling them like there was no tomorrow. But, in all seriousness, the Mayan calendar is important to contemplate even now, but for reasons entirely different from those the mass media were (mis)representing. The calendar (actually calendars, plural) are a depiction of the evolution of human consciousness in direct relationship with the larger waves of consciousness (sometimes called "worlds" or "underworlds") that originate from Mother Earth and, ultimately, from the greater cosmos.

The Mayan calendars never prognosticated the end of time, but rather *the beginning of the end of the way we think of time.* Our misreading of these calendars comes from our attachment to material forms. We focus on the things, the dates, the points in an imagined linear timeline instead of the overall cycle. We saw a so-called end to the Mayan calendars because of our addiction to linear historical consciousness.

The Mayan calendars are fundamentally about cycles of time and corresponding unfolding energies. They do not depict time as a line with a beginning and an ending. Rather, they mark the small and grand movements of nature and the corresponding unfoldings of consciousness within nature and within each of us.

Much was made of the "Mayan Long Count," because most anthropologists promoted the idea that the Count began at the dawn of written civilization, or the dawn of "history," around August 11, 3114 BCE, and ended on December 21, 2012. However, in my opinion, the researchers who best comprehend the Mayan concept of time understand that the calendars could not really end, nor even begin, at a particular point in time; rather, the calendars can be extrapolated back before Western scientists posit the big bang, and forward indefinitely into the remote future.[6] How could it be otherwise when the calendars are based on cycles of time, not on time as a line?

For purposes of this book, the Mayan calendars are valuable not because of linear dates they may or may not correspond to, but because they are a part of ancient sacred knowledge that expresses the relationship between the energies of nature and consciousness. In the West, with the development of abstract notions of time, we moved away from direct engagement with nature. But part of our Western history is also this ancient original thinking as expressed in the Mayan calendars and the other Indigenous knowledge that originated in the Americas.[7] We should remember this when we talk of the exportation of global culture from the West. Western culture has its own sacred roots, and this part of our heritage can be enormously enriching to explore.

The Mayan calendars are important because they acknowledge and formalize the unfolding processes of creation. This is particularly encapsulated in the Tzolkin Calendar's (otherwise known as the Sacred Calendar) count of days (a 260-day period). In the Tzolkin count, the numbers 1 through 13 are known as a *trecena,* which is the Spanish word for what

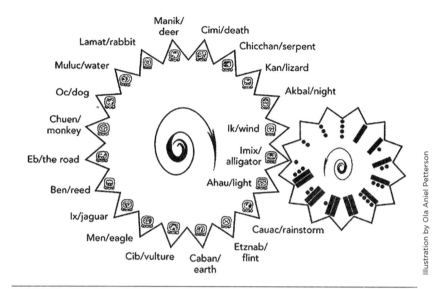

Figure 4: Tzol'kin - Two rotating cogwheels describing how the 260 energies (Tzol'kin combinations, here 1 Imix) are generated and interacting by the twenty day-signs and the Thirteen numbers.

in English might be called a "thirtnight"—although there is no English word for a series of thirteen. The 13-day count is said to be a reflection of the unfolding of evolution, from seed to mature fruit, which takes place in thirteen steps.[8] Each day is assigned one of twenty different signs. The signs are counted in a specific order, resulting in a 20-day cycle called a *uinal*. The two counts, the trecena and the uinal, run concurrently, so each day is characterized by both a number and a sign. This is commonly illustrated as two cogwheels, each moving one step a day to produce a new combination.

Because there are 13 by 20 different possible combinations, adding up to the 260-day count, every day in the count has its own specific energy, or vibration. The vibration occurring on one's birthday is considered significant, as it contains the seed of one's life purpose. Likewise, Mayan "Daykeepers" (diviners) interpret each day's energy to help guide appropriate action.

The Mayan calendars are still relevant today because they tell us we are on the precipice of an important transition from a tumultuous age to at least the beginning of a more harmonious, if not golden, age. But this is not something the Mayans alone foresaw. There are many calendars, all converging on the current time and all marking this as a time of great transition. Like the Mayan calendars, these calendars are also often mistaken to signal the end times. But it was never the end; it is only another beginning. We are awakening to a new form of intuitive awareness, as prophesied in Hopi, Navajo, and Inca stories, as well as Egyptian, Babylonian, and Hindu stories, among others.

These various prophecies have all foreseen this as a time when consciousness unfolds to a new level of awareness, or a new world(view). The Incan prophecy of the Eagle and the Condor, in which the eagle of the north and the condor of the south fly together again after centuries of separation, represents the rebalancing of the masculine and the feminine, the mental and the heart/intuitive, the old and the new. This time of rebalancing is not an end of time, nor is it the pinnacle of evolutionary progress. It is only an apocalypse in the original meaning of the word: an "unveiling" or "revelation."[9] It is an unfolding of human consciousness in keeping with the unfolding of planetary consciousness. There is more to the story that will come long after this era passes.

Why Are the Cycles of Consciousness Important?

When we do not think in terms of a large enough cycle of time, it is impossible to understand the big picture and to make informed decisions, particularly with regard to complex phenomena such as education, international politics, or global climate change. We tend to look at any period of history through the lens of modern thinking and see only those aspects of the past that validate the current cultural worldview.

Underlying this misconception is the core belief that time is a line, and that whatever occurs now has been built upon the past body of knowledge. This belief is held not only by the scientific community, but also by the society at large. Whatever is old is considered inferior. The belief has become so extreme that we tend to believe the results of a study released today more than the results of a study released five years ago or even five days ago. In fact, a recent study by a news station showed that whenever they begin a story with the phrase "A recent study showed," people will believe whatever comes next. I am joking of course, but the question remains whether you started to believe me.

In short, we tend to believe that we are building upon the past in linear increments, and therefore continually improving, always getting smarter and better. If only this were true! It is difficult to see that what is occurring today is merely the blossoming of seeds planted long before, emerging in their proper season within a greater cycle. The New Age is not new. For example, today we are rediscovering wholistic approaches to healing that were once the norm. We are again placing importance on the environment in which we are treated, and increasingly demanding to be seen as whole human beings.

As these changes occur, many of us cling to the idea of linear progress, believing that what is newest is best, and we consider these wholistic approaches a new development, newer than mainstream medicine. If asked what we think of medical care long ago, for instance in the Middle Ages, we reflexively downgrade that era as backward, barbaric, unclean, or inferior, and probably dangerous. It would surprise many of us to learn that in the Middle Ages, patients were not placed in the soulless box of a hospital room; instead, they were housed in a sanatorium, which was a beautiful building with grounds providing plenty of fresh air, sunlight, plants, and soil.

Back then, it was understood that nurturing the soul was essential for healing. People were more likely to see their lives in a balanced and whole (holy) way, integral to the unfolding processes of nature. We are actually coming back to this view now, though in a somewhat different way. I recently saw a newspaper story heralding new hospitals now being built with plenty of sunlight and fresh air. Many of us do not associate any of today's changes with an era that came before; nonetheless, this is frequently the case.

What Needs to Come Back into Balance?

In the West, it is the relationship with time that is most unbalanced; in fact, we are addicted to the idea of linear time. Think about people you know. Many are narrowly focused on the future, while others remain stuck in the past. If you are one of those fixated on the future, you typically think that things will be better later, when one thing or another changes—when we get a new job or quit our current job, when we finish school or go back to school, when we get a new car, or when we move to a new neighborhood. In short, there are innumerable pending solutions to our present circumstances, all in the future. We have trouble accepting what is right now as a blessing.

The same is true for people who tend to dwell in the past. When this is the case, our recurring thought patterns tell the body that the event is occurring in the present, and it truly is for as long as we are preoccupied with it.

There is nothing wrong with planning for the future or occasionally taking stock of the past before moving onward. But when we dwell in the past or become fixated upon the future, we become unbalanced and can literally make ourselves time-sick.

Time-sickness is an artifact of the confusion about time. I believe it is actually a form of homesickness, because what we are missing, without realizing it, is a time when we lived in continuous connection with our homeland. We still yearn for that place of origin as a source of renewal and regeneration.

I have never heard of someone having a problem because they were too much in the present. But it is too simple to just say, "Live in the present," because the present is so much more than what is implied in the English tense. The present is not merely a point in time between past and future. The present is an open doorway to all the confluent rivers of time.

In one of the early dialogues, Grandfather Leon spoke about how in the Navajo language, "We have to go back in order to begin."[10] He explained how in Navajo, you go from the present time to an earlier time (which he called "alpha time") and then to the origin, where you can access "original instructions" for how to live on Earth that are appropriate at any time. The three levels of Navajo language allow native speakers to move back and forth fluidly into any time—and may even allow for what is called the future to be seen as easily as the past. The best way to picture this might be to imagine surfing a wave. When you catch the wave just right, you are surfing the flux of the central origin, which is continually being created by the past and future of the waves that feed into it. You do not just go backward or forward in linear time; instead, you are riding the flux of time.

Chapter 4

It's About Time

What Is Time?

Although I am asserting that linear time is an illusion, time itself is not an illusion—for without time, there would be no movement, no activity. So what, exactly, is time?

One of the most intriguing statements I ever heard in the dialogues came from Grandfather Leon, who said, "Time is the fifth element." I first heard him say this in one of the earlier dialogues, and he repeated it throughout the years until slowly something of what he meant dawned on me. The first four elements are fire (or as Grandfather Leon usually said, "light"), air, water, and earth. He often referred to the elements themselves as the Creators, which he did not intend as a metaphor or abstraction because the elements do create and sustain life. But what did it mean to say that time is the fifth element?

Then, one day, I saw. If time is the fifth element, then time is the animating force that interconnects all the elements.[1] In other words, *time is spirit*. And if time is spirit, then it is not something that exists outside, independent of us, but something within us, an ever-present source of nourishment. It is (at least potentially) a wellspring of creativity infusing our lives with meaning. If time is spirit, then time is a part of us, for we are made of the elements. And if time is spirit then, like air, we are also immersed in it. We must have all the time in the world!

If time is spirit, then we do not need to act in haste, but can allow the processes of time to unfold, the way a spider unfolds her web from the center of her being. If we, on the other hand, spend our lives rushing around because we treat time like money, we are inflicting pain upon our spiritual self. As Larry Dossey pointed out in his classic book, *Space, Time, and Medicine,* many illnesses occur because people are time-sick; they have an unhealthy relationship with time. But when we move in rhythm with the changes in nature, designing our behavior to fit within those existing flows of energy rather than trying to willfully control them, not only do we harmonize with our surroundings; we synchronize.

And we get well.

The elements are not truly separate; they are primal aspects of the whole that always work together. A volcanic eruption, for example, demonstrates the interconnection of fire, air, water, and earth. From the womb of Mother Earth, out flows lava, a river of liquid fire that, like all rivers, flows toward the ocean. Upon contacting the ocean, the water cools the fire and creates new earth right before our eyes. I have seen this happening on the Big Island of Hawaii—getting close enough to have my shoes nearly burn—and it is one of the most awe-inspiring things you could imagine. But we need not look for such dramatic examples to see the interconnection of the elements. School children observe the interconnection of the elements when they conduct experiments heating water to create steam, freezing water to create ice, or growing a seedling in a tiny pot. A seed cannot become a plant or tree by itself. It needs earth, water, air, and light to grow. The elements all interpenetrate to create life; this is their inherent beauty. If time is the fifth element, then it must follow that the other four elements act in concert with time for life to occur.

If we can begin to think of time in this way—as an element—I see enormous potential for healing the split consciousness we have been saddled with ever since we began to think of time as linear. The by-product of linear thinking is separation and alienation from life as it organically unfolds, but when we are immersed in time as spirit, we feel connected, blessed, and creative.

Time is our relation; it does not exist apart from us, but in relationship with us and with all of life. When we establish this kind of familial relationship with time, it no longer seems possible to think of it as a line moving away from us, nor as something separate or abstract. The linear way of thinking creates distance, but the relational way creates intimacy. Time is an essential element in the interconnectedness of life. If we think of time this way, it can transform our lives.

I = Time

Little Bear began a dialogue on time and origin one year with a provocative statement, "I am time. I equals time." He then stated that if I equals time, he wanted a divorce from time and posed this question, "If time is no longer a part of me, now what?" As he does with many of his kick-start questions, Little Bear was already turning its meaning around, even as participants considered the opening question. Before he asked the question, I doubt many had ever before considered time to be a part of them. Little Bear formulated the question as a statement, making the notion a given. His opening also wonderfully encapsulated the last five hundred years of human history in a simple equation.

Going back to what is called prehistoric times, or what anthropologists call mythical time, time was a part of us and we were a part of time. We had not yet separated ourselves from the natural rhythms of the earth, but lived attuned to the earth's rhythms. Our foremost teacher was nature herself. When we deviated from her rhythms, we learned ceremonies from nature to bring us back into balance and harmony.

So what happened to change this? We began to believe that we could harness, even defy, nature's rhythms for ourselves and control our future through willful human acts. This was the beginning of a separation of our consciousness from nature, which culminated in the reification of linear perspective in art. As noted earlier, perspective portrays time in terms of sight lines projected away from a stationary single eye toward the future. This is both *eye equals time* and *I equals time,* because the conditions were created for the egoic perspective to supplant nature's rhythms. Time became an abstraction, and abstract time allowed us to imagine that we did not need to adapt to nature anymore.

We began to invent ways to outwit nature. With artificial light, we can stay up as long as we like and get up whenever we want. We plant what we want, when and where we choose, by continually adding soil fertilizers and eradicating pests chemically. We eat foods in or out of season, transporting them to supermarkets from all over the world. We even choose where to live in complete defiance of common sense, whether on top of cliffs or next to riverbanks, and then wonder why we suffer landslides and flood damage.

The reality is that every time we manipulate nature's rhythms, we create unintended consequences that then require us to make still further changes. When there is flooding, we dam rivers, and when the fish cannot make it through, we worry about that problem later. When there is a landslide, earthquake, or tsunami, we rebuild in the same place and just double down the reinforcement. We rarely step back long enough to reconsider our position in a way that works *with* instead of *against* nature. We continually make new decisions that are not well thought out, when we could learn from the perennial wisdom that exists everywhere on the planet and integrate that wisdom into our thinking. In effect, we could think originally for a change.

Those who have bravely spoken up against the foolhardy disregard for nature have been branded as being behind the times or standing in the way of progress. Fortunately, increasing numbers of people are now questioning the way we have been living—including the notion of time, for at the root of the difficulties is the way we have come to misunderstand and misuse time.

As Little Bear pointed out in the dialogue, we make so many decisions based on how much time we do or do not have, how much time we waste, and so forth. Until now, we never questioned why we make decisions this way, because linear time was not up for debate; it has been a hidden part of our infrastructure. Time remains a part of us, but now it is linear, abstract time—not the rhythm of nature—that has become embedded in the way we live. And, because abstract time is so much a part of our lives, we have had difficulty seeing it.

I am heartened, however, to see this finally come back into awareness. We are beginning to rediscover the difference between natural cycles of time and clock time and all of us, Indigenous and non-Indigenous, are awakening to the colonizing effect that clock time has on our souls. We do not want clocks to run our lives anymore.

Like Little Bear, many of us want a divorce from clock time. And, essentially, we have been engaged in a messy divorce from this sense of time throughout the last century. This is why the French impressionist painters were such a breath of fresh air. Impressionism allowed us to reexperience the moving essence of nature. When we look at Monet's lilies, we sense the gentle rocking of the lily pad atop the water and may even feel the breeze sweeping through our own bodies. We reconnect with the spirit of place when we experience the works of the Impressionists—more so than the static paintings in the school of "realism."

In a different but related way, Picasso turned our heads around; he confused our rational mind, but engaged our intuition. We were forced to let go of our habitual way of seeing in order to reintegrate the essence of a human figure. A Picasso figure may be an ostensibly fragmented image, but an underlying wholeness comes through, animating and breathing life into its form.

If nothing else, Picasso liberated us from a rigid attachment to geographic time and space to play with new (old) sensations, which had been repressed by linear time and realism. By breaking out of the limited way we have been seeing geographic space, we also break out of our fixation with abstract linear time. Breaking away from clock time is that much more liberating because we are freed from the cause of much of the stress in our lives.

What Is the Original Purpose of Stress?

In his book *Creative Stress*, James O'Dea explores the idea that stress is a natural part of life, and that we need to engage our stress rather than eliminate it. He says it is too simple to say that stress is bad and relaxation is good, for stress can be a great teacher. He makes an excellent point.

The original purpose of stress is to heighten our alertness to changes in the environment that can affect our well-being. We are meant to listen to these signals from nature. O'Dea reminds us that stress teaches us to recognize what is true. When we ignore signs of stress, we do not eliminate stress. All attempts to keep stress at bay—whether self-medicating with alcohol, drugs, or television, for example—postpone and ultimately exacerbate stress. It is imperative that we face our stress and its causes, rather than reflexively running away.

The modern approach to leisure has become one of the sicknesses of our age, because we have created an artificial dichotomy between work and play. We crave leisure because, at some level, we recognize that leisure means being in tune with natural rhythms. But we have so distanced ourselves from natural rhythms that we have forgotten how to reconnect with nature in our everyday lives. We have come to depend on vacation time as our savior, the time when we can unwind our compressed lives and regain our sense of wholeness for a time. *Unwinding* is indeed the appropriate metaphor for those who live by clock time, for we have become so wound up that we are like alarm clocks, poised to go off at any time. Unfortunately, when our vacation finally comes and we arrive at our exotic locale—that is, a place existing in non-Western time—it takes us so long to unwind that, by the time we feel restored, it is time to return to our frenetic schedules.

Much of our stress today is the result of our unnatural relationship with time. We have developed a time disease that comes from our unwillingness to see our lives as emerging from, and held within, the rhythms of nature. We want to live outside of natural time in our own imagined time, an imagined abstract world of thought. Much of our stress also occurs because we think too much, as Matthew Bronson pointed out in one of our dialogues: "Time disease and thinking-too-much disease are the same disease," he opined. And indeed they both promote the same function, which is to divide the world into fragmented pieces rather than to see the world in wholes.

How Can We Reframe Our Stress and Anxiety over Time in Our Everyday Lives?

A consequence of seeing time as linear is that we see the world as divided into discrete events or separate parts. This creates a fragmented consciousness that eventually leads to stress and anxiety. The problem is not only that we see the world as divided, but also that we are convinced that this is the literal truth. But the world is not inherently divided; it is in essence an undivided whole. It is possible to recognize this only when we change the way we see time.

If we think of time as an element, connected with the other elements in wholeness, the way we relate to time changes. Time now becomes an ever-present source of sustenance and renewal instead of the source of limitation and stress that it is for many of us now. Our life becomes whole instead of divided. Segmented activities that compete for our attention are replaced by life activities that complement and reinforce the whole.

A life that begins one place and then continually diverges from its origins is unsatisfying—there is always a feeling of something missing. A life that is organized around a coherent center is satisfying, because it feels whole. When we organize our life around a coherent center, all of our activities become ripples of time that flow from the same center. This can happen in time—literally—if we allow time itself to be the central organizing principle in our lives. With time in the center, any division in the various aspects of our lives disappears because time is the impetus that generates interconnectivity among the various aspects of our lives. Peace and creativity then flow naturally from that central nexus.

When we understand time as an unfolding, we know when it is the right time to act. This is different from, but does not preclude, acting when appropriate within the limits of linear time. For example, Grandfather Leon was once supposed to give an hour-long talk at a conference I organized. He arrived forty-five minutes late (by clock time) and asked me how much time he had left to give his talk. I reluctantly told him he

Figure 5: Time as coherent life purpose placed in center.
Ripples of time (supporting activities) surround center
and interconnect the events in our lives.

had only fifteen minutes because another activity that could not be moved was scheduled to begin next. He nodded and then began his talk, speaking more slowly than I had ever heard him speak before. I soon realized what he was doing. He was stretching time by allowing the pauses between his words to carry his deeper message. He spoke for exactly fifteen minutes but still conveyed everything he would have if he had been able to speak for a full hour.

Many people I admire have discovered how to walk in balance by living in harmony with nature's timing, continually attuning to and being nourished by spirit (natural timing) in spontaneous ways. When we do this, we let go of all our preoccupations with clock time (and abstract knowledge) and organize instead from a central place of nothing, like the eye of a hurricane. Mahayana Buddhists call this *sunyata*. Carlos Castaneda described it as "erasing personal history." Taoists might call this *Tao*, living in harmony with the natural order, which can be lived but never adequately described. For a Navajo, this would be "living the order" or walking the sacred path of the Beauty Way (*sa'ah naaghai bik'eh hozhoon*).[2] All of these are different expressions of original blessing, of living creatively in nature's time rather than being on time in clock terms.

Is It Possible to Come Up with an Original Thought?

I have given the long answer. Now, I will attempt the short answer, which is much harder to do. There are, as I mentioned at the outset, hundreds of ways of thinking about this question, but in the end, it comes down to two. Either we believe, as the Laguna tribe does, that only Tse che nako ("Creator," "Old Spider Woman," or "Thought Woman")[3] is capable of original thought and that all of us are merely borrowing her thoughts, or we believe that each of us has the capacity, even if it lies dormant, to directly access Origin at any time. Because I do not like dualities, I will choose a third way, a way in which both are true.

As Paula Gunn Allen told us in the dialogues, it is customary in Pueblo pottery to begin a new pot with the shards of an old pot. This means that broken fragments are reused to make wholeness. It also means, if followed religiously, that every pot made is directly connected to every other pot that has ever been made by the tribe. This custom captures the essence of original thinking.

Original thinking is a remembering or a putting back together of something that has always existed, but has been torn apart. Original thinking embraces the wholeness of time; it is both new and old.

Original thinking comes from a place of timeless origin—a place of original relationships and alliances—and it moves us toward a world that is increasingly inclusive, interconnected, and whole. It is not either/or thinking, but both/and thinking. It chooses from the whole landscape of possibility, from both ancient wisdom and modern innovation, weaving both into a wholeness appropriate to the situation at hand. It is not a fixed way of thinking but a process.

To think originally is to continually open ourselves to the widest window of consciousness that we can imagine. It is beyond my power to know how wide that window can open. I only know that to the extent I am able to open my own consciousness, I increasingly lose myself in abundant creativity and abiding joy. And, because of this, I know something else. It is possible to lose oneself and gain an identity. It does not seem to make sense, but it is so. For, in losing yourself, you find your relationship with everything surrounding you, and that is the source of a new and much more fluid identity. That is as close as I have come to origin.

I also suspect that to truly think originally is not to think. "The Tao that can be told is not the eternal Tao." This is all I will say on the question for now.

PART TWO

Departure: What Does It Mean to Be Human?

Many of us in this time have lost the inner substance of our lives and have forgotten to give praise and remember the sacredness of all life. But in spite of the forgetting, there is still a part of us that is deep and intimate with the world. We remember it by feel. We experience it as a murmur in the night, a longing and restlessness we can't name, a yearning that tugs at us. For it is only recently, in earth time, that the severing of the connections between people and land has taken place. Something in our human blood is still searching for it, still listening, still remembering.

—Linda Hogan

The historical mission of our times is to reinvent the human—at the species level, with critical reflection, within the community of life-systems ... by means of story and shared dream experience.

—Thomas Berry

When I pick up a person from the street, hungry, I give him a plate of rice, a piece of bread. But a person who is shut out, who feels unwanted, unloved, terrified, the person who has been thrown out of society—that spiritual poverty is much harder to overcome.

—Mother Teresa

A youngster held a beautiful bird behind his back. Trying to embarrass a respected elder, he said, "There's a bird, is it alive or dead?" If he replied, "it is alive," the youngster would kill the bird by squeezing its neck to prove the elder wrong. If he replied, "it is dead," the youngster would release the bird to fly away and again prove the elder wrong. When the youngster repeated three times, "There's a bird, is it alive or dead?," the elder thought deeply and replied, "It is in your hands."

—A story from India

The Journey of Scar Face

A Blackfoot Story

Scar Face lived with his grandmother because his mother and father had died shortly after his birth. His face had a birthmark that set him apart from all others and became the source of ridicule and shame. Because he was different, he was taunted by the children and other members in the tribe. As Scar Face grew older he withdrew and spent much of his time alone in the forest befriending and learning the ways of the animals he encountered. It is said that he learned to speak with them. And through them he learned how to be related with all things.

As Scar Face grew older he fell in love, as young boys do, when they come of that age. The focus of Scar Face's affection was a young woman, Singing Rain, the chief's daughter. Singing Rain was also a special person, kind and with a gift of insight. Although all the other young men competed for her affection, it was Scar Face who she came to respect and love because of his honesty and good heart. However, when Scar Face asked for her to marry, she revealed her sacred vow to the Sun never to marry. This was her pledge of spiritual piety in the way of the Blackfoot. The only way she could marry was if the Sun were to release her from her pledge. And so, it is said that Scar Face began his visionary journey to the land of the Star People.

Scar Face did not know where the Star People lived, only that they must live in the direction the Sun set every evening, beyond the Great Water in the West. So Scar Face prepared himself with help from his grandmother, and when he was ready he set forth on his journey, a journey to the land of spirit. He first traveled familiar territory, but then began to enter into lands that neither he nor other members of his tribe had ever seen.

As the snow of Winter began to fall, a hundred paths became open to him, and he became confused; he did not know which way to go. He met a wolf on one path, and with great humility asked for help and direction. Knowing the goodness of his heart, the wolf spoke to him and guided him to the right path. He traveled that path for a great distance until he came to another series of paths. Again, he became confused. He stopped, set his camp and prayed. Soon a mother Bear and her cubs appeared on the path in front of him. Again, with great humility he asked for guidance from the mother Bear. The Bear spoke with great kindness and pointed out to him the right path. Scar Face followed the bear's path for many days until the path ended. Now there were no longer any paths in front of him to follow, only the vast expanse of the great forest. As he stood and pondered in front of the forest, a wolverine approached him. He called out, "Good wolverine, my friend, I need your help." Again, he asked for direction and help from this friend. Knowing his heart and the nobility of his quest, wolverine responded with great kindness and guided him through the forest to the edge of the Great Water, where, exhausted, he made camp. He thanked the wolverine, as he had thanked each of the animals that had helped him, by offering them a gift of song and tobacco. He could see a twinkling of lights across the Great Water, and he knew that was the land of the Star People.

Scar Face did not know how to cross the water to "that place that his people talked about." But he was determined to find a way. Then two snow geese swam by and offered to take him across the Great Water. When they arrived on the other side, he thanked the geese in the same manner as he had done the other animals for their kindness and great service to him. He made camp and then fasted and prayed for three days and nights. On the fourth day, a path of sunlight began to form in front of him leading toward "that place." He leaped onto the path and followed it as it took him higher and higher into the sky. When he reached the end of this path of sunlight, he came to a beautiful forest and another path that was of great width as if it had been made by thousands of people traveling on it for a long, long time. As he followed the path he came upon a quiver of arrows leaning against a tree. He wondered whom they must belong to, so he waited to see. Soon, on the path coming from the other direction was an extraordinary looking Warrior dressed in richly decorated white buckskin. As the Warrior approached, Scar Face could see that this man was an image of perfection. He asked Scar Face if he had seen a quiver of arrows. In response, Scar Face showed him where the arrows were. Grateful and curious, the stranger introduced himself. "I am Morning Star." Then he asked Scar Face his name and where he was going. "I am called Scar Face, and I seek the lodge of the Sun."

"Then come with me, Sun is my father and I live with my mother Moon in his lodge."

When Scar Face arrived at the lodge of the Sun, he saw that the walls were painted with the history of all people of the world. Morning Star introduced Scar Face to his mother the Moon. As his father the Sun entered the lodge, a great light entered with him. Morning Star introduced Scar Face to his father Sun, the greatest chief. Scar Face was so impressed that he could not bring himself to reveal his reasons for coming to the land of the Star People. Sun and Moon treated Scar Face with great hospitality and asked Scar Face to stay with them as long as he wished. Over the next few days, Morning Star showed Scar Face the many paths in the beautiful land of the Star People. There was one path to a distant mountain that Sun had warned Morning Star and Scar Face never to go near. It was a mountain on the top of which lived a flock of seven giant birds that the Star People greatly feared.

One morning, Scar Face woke to find Morning Star gone. Scar Face arose and quietly left the lodge of the Sun to take a walk and decide how he might ask Sun to release Singing Rain from her vow. He thought he might meet Morning Star and ask him for advice. But as he walked, he began to feel that something was wrong, and the nearer he came to the mountain where the Giant Birds lived the stronger his feeling became. He knew that there was some reason Morning Star had gone to the forbidden mountain.

Scar Face set out in search of Morning Star. As he climbed to the top of the mountain of the Great Birds, he found Morning Star engaged in a ferocious battle with the birds. These birds were indeed savage and extremely large. They were about to overcome Morning Star when Scar Face joined the battle. Scar Face fought valiantly and soon turned the tide of battle. One by one, Scar Face and Morning Star began to kill the Giant Birds until all seven were slain and their tail feathers taken by the two warriors.

Tired, yet proud of their accomplishment, Scar Face and Morning Star descended the mountain and returned to the Sun Lodge to inform Sun and Moon of the defeat of the Star People's most feared enemies. Sun and Moon were very impressed by the courage shown by both young men and were especially grateful to Scar Face for saving the life of Morning Star. In honor of the courage of Scar Face, Sun offered to fulfill any desire he would request.

Yet, Scar Face could not speak his greatest desire. He remained silent until Moon, knowing his heart, spoke of Scar Face's love for Singing Rain and her vow to the Sun that prevented them from being together. Sun immediately responded by saying to Scar Face that he would release Singing Rain from her vow. Sun touched the cheek of Scar Face, and the scar he had borne all his life disappeared. Morning Star in turn gave him special personal gifts and revealed to him that he was his spirit father, confirming the feeling that Scar Face had all along. Then Sun and Moon began to sing songs in praise of Scar Face and Morning Star. Sun and Moon then gave Scar Face many gifts. In addition, Sun renamed Scar Face "Mistaken Morning Star" because now without the scar on his face he looked like Morning Star. Sun taught Mistaken Morning Star his own special dance, the Sun Dance. He said that if Earth People wished to honor him and bring health and well-being to their tribe, they should dance the Sun Dance each year when he has reached the highest place in the sky. Then Morning Star led his Earth son to the path called the Wolf's trail (The Milky Way) and placed a wreath of juniper on his head. In an instant, Mistaken Morning Star was back on Earth and on a path leading to his own village.

Singing Rain was the first to meet Mistaken Morning Star as he approached the village. He told her that Sun had released her from her vow, and she knew in her heart from seeing and feeling the magnificence of him that they could now be together always. Mistaken Morning Star called the people together and taught them the rituals of the Sun Dance. He showed the women how to build the Sun Dance Lodge, and he taught the men how to conduct the sweat lodge ceremony and raise the Sun Dance pole. He taught them about the nature of sacred visioning. He taught them from "that place that the Indians talk about."[4]

Chapter 5

Purpose, Potential, and Responsibility of Being Human in the World

What Does It Mean to Be Human?

Why were we born? How do we fit into the greater cosmos? What is our purpose here on Earth? What can we do, individually or collectively, to better this world? These questions have no answers that are correct for all persons, yet they are compelling, and many of us return to them, again and again. But the answers are elusive, sometimes maddeningly so.

I feel most human, ironically, when I let go of thought and lose myself in an activity. I could be praying, singing and chanting, or immersed in meditation. But this is not necessary. I could also be walking with my dog in the foothills of the mountains and listening to the familiar songs of the local birds—or I could be simply sitting on the back porch, enjoying the air and watching the clouds move on. I do these things almost every day, but each time is special—a renewal of spirit.

A cloud is a symbol of thought—the best one I know. I delight in seeing a cloud form in empty space—something that often happens over a mountain. What is it about the mountain that attracts the cloud? Watching a cloudless sky reminds me of the need to empty myself in order for new thoughts to form. Thoughts gather gradually, much like clouds, and if I nurture them correctly, they eventually bring a downpour of inspiration.

The kind of thinking that is most nourishing to me occurs when thoughts come to and through me in an effortless rhythm, just like breathing. A good thought can spawn another good thought, but I have to guard against holding onto any one thought, for thoughts can be addictive.

When I am paying attention, I breathe in a thought, enjoy it for a moment as if tasting a fine wine, and then let it go. I let the thought die a peaceful death. I have found that allowing ideas to pass is the surest way to original inspiration. It is a matter of trust, in the same way that breathing out trusts the next in-breath to come. But the thought that comes in is not my own; it is from Nature. That is why we say a thought *comes to us*. Sometimes, if only for a precious moment, I get a sense of what it is to be human. That momentary awareness is a glimpse into something immense, and I cannot articulate it. The only thing I know for sure is that I am a living, breathing, and thinking being in a vast cosmos that is also living, breathing, and thinking.

Clearly, thought has a lot to do with being human—but not in the way we moderns tend to assume. Modernity has seized upon one particular type of thinking and equated it with essential humanness. It is our capacity for rational thought, we have determined, that defines us—both separating us from, and elevating us above, other animals and all other creatures. But what if, as we have been implying, this belief is simply wrong? What if we have been barking up the wrong tree all this time?

Rational thinking could be fool's gold—the glitter of something we mistake for more than it is. We may misconstrue the nature of thought entirely. Our thoughts may not be our thoughts at all, but a filter of Great Spirit's thought, an inlet to a greater knowing that extends far beyond the reach of rational comprehension.

I believe the potential of the human being emerges out of the potential of all of creation, but that we regularly underestimate both the power of creation and, by extension, our own power, for we are a microcosm of the macrocosm. Our thoughts and abilities are not greater than the rest of creation, but because of it; whatever intelligence we have is because creation itself is intelligent, and it is arrogant to assume otherwise.

At the same time, the human being is potentially far more than we think ourselves to be. Like a caterpillar about to become a butterfly, we identify with our current body (of knowledge) and our current way of thinking, completely unaware of the potential that lies within us and that, unbeknownst to us, is inexorably unfolding and ultimately transforming us into something unimaginably magnificent and beautiful.

How can we awaken from the sleep of modernity and open ourselves to our full potential? A sense of urgency, or emergency, is necessary to precipitate our *emergence* from our cocoon of ignorance. This is the stage where humanity now finds itself, because the actions of human beings—actions that have come about because we mistakenly think we are separate from each other and from the rest of creation—have put our lives in peril, along with the lives of countless other species. We have to evolve now or perish. And to evolve, we have to change the way we think—and the way we think about thinking.

So what is stopping us? It is our over-identification with our intellect—and with a particular kind of thinking, which is really a degraded form of rationality—that holds us back. We are so addicted to this way of modern rationality and believe in it so much that we fail to recognize the problems it causes and try to overcome these problems with the same kind of thinking that got us into the mess. But that is the biggest mistake of all.

How did this situation come about? It is because of a peculiar tendency of modern rationality—a tendency to first dissect reality into parts and then to represent these parts as if they were the whole. In a similar way, we have dissected the human being into parts, and then latched onto rational thinking as the most important or most unique aspect of being human, when it is far from it.

The preeminence of rational thinking has become the single greatest delusion of modernity. We are not entirely—nor even primarily—rational, nor should we be. We are also not only "spiritual beings in a physical body," as is sometimes claimed, for that characterization, too, represents the part as the whole. No, we are not reducible to any singular aspect of our being. Our true humanness comes from the totality of our diverse but interconnected aspects: our spiritual, intuitive, mental, emotional, and physical selves. None of these attributes should be discounted from the whole human being.

This includes our physical nature; for as long as humans have bodies, our animal instincts will continue to be important. Even when we have epiphanies that appear spiritual and/or intellectual, this does not eclipse our animal instincts. In the end, there is no separation between human and animal. We are a great many things, and we have enormous untapped potential, but we will always remain an animal.

As a young boy, I once had a rude awakening that enabled me to see this clearly. At the time, I was blessed to be living on the north shore of Long Island, New York, immediately adjacent to an extensive area of woods that led to a large body of water (Long Island Sound) off a so-called dead end street. In actuality, the dead end was teeming with life just outside its borders. I used to play in the woods quite a bit, and through my father's guidance and some of my own naïve explorations, I learned a little about the squirrels, birds, ants, bees, and frogs and, on occasion, skunk, possum, and other wildlife.

One day, when I was playing in the woods, I heard a crashing noise as something was rapidly approaching. I turned to see what animal it could be and was surprised to see it was a neighbor boy, Joey,[1] running at breakneck speed, and from quite a distance. I stayed motionless, and he did not see me as he came closer and then rushed on by. Then, suddenly, he stopped, dropped his pants, and proceeded to take a bowel movement right there in the woods—and then, just as fast, he pulled his pants up and ran on. It was then that I realized that all the stuff I had been hearing about animals and humans, supposedly different creatures, was nonsense. We were animals too.

It was soon after that I read Desmond Morris's *The Naked Ape*. It was 1967. I was twelve years old, and the book made a vivid impression on me. Morris revealed how much of our animal nature we were either covering up or blithely unaware of. How many women consider, as they apply their daily makeup, that the wearing of lipstick and rouge is imitating their natural state of sexual arousal? Not many, I suppose; nor do men, who are mostly clueless as to the root causes of their stimulation. We do not think about these things because we think we have evolved beyond them—that we have them literally under wraps. Yet, even as we wear the most civilized of costumes, be they ballroom gowns or tuxedos, we cannot cover the fact that underneath it we are women and men with the same bodily needs as all animals. To state the obvious, we need water, food, and a place to sleep, urinate, and defecate, but we also need affection and some form of social, if not sexual, relationships. We act more like other animals than we realize—for instance, when we court our potential mate, curry favor with our superiors to climb a business or social ladder, dominate a conversation, or lovingly embrace a child.

Animals have also been our teachers. For millennia, humans learned a great deal from other animals—and not through laboratory testing. We observed animals in their natural environment: what they ate, how they ate, what they ate when they were sick, how they hunted, how they defended themselves, and a myriad of other skills and capacities. Many healing herbs, such as osha (bear root), echinacea, and yarrow were discovered by watching animals eating or using them to treat wounds or illness. We modeled much of our healing ways and practices on what the animals did.

We also modeled much of our arts, sciences, and religions after the patterns of nature, even if we are unaware of it. There is great beauty in nature that we sought to imitate. Black Elk spoke of birds building their nests in circles, "for theirs is the same religion as ours."[2] But all of humanity took their original inspiration from something in nature. As Buckminster Fuller understood, we cannot create something that is not already part of universal design. Ultimately, humans are imitators, or should be. The emerging field of biomimicry seeks to do just that, looking to the natural world for help with human problems.[3]

In modernity, however, we have forgotten our connection to other animals, plants, and minerals, and have sought to elevate ourselves above the rest of creation. One of the ways we separate ourselves from the animal kingdom is by distinguishing between human intelligence and instinct, relegating the latter to a lower level. But is this really the case? Members of the animal kingdom display great achievements, such as the complex social communication of bees or ants. When bees perform a waggle dance, for instance, we find it a fascinating curiosity, but "It isn't thought!" we protest; it is merely instinct. Never mind that the dance informs the rest of the colony as to the location of water, nectar, or pollen. This could be considered an act of great intelligence, but is instead dismissed as inferior, mainly because we do not understand it.

What Is Instinct?

What we dismiss as mere instinct in other creatures is actually one of the most important human capacities—something we not only share with the rest of creation but that we rely upon even today. We know it as a gut feel or sixth sense. I consider instinct to be an aspect of original thinking because it emerges out of a ground of consciousness that connects us to the source of our existence and our original purpose for being. Instinct shows us how to behave in a bodily congruent way that is in tune with our surroundings. But instinct is not limited to the body. A gut instinct, for instance, can help us make the right intellectual decision, for instinct does not annihilate mental cognition; it is merely a harbinger of an intellectual understanding to follow. As Emerson wisely said:

> You have first an instinct, then an opinion, then a knowledge, as the plant has root, bud and fruit. Trust your instinct to the end, though you can render no reason. It is vain to hurry it. By trusting it to the end, it shall ripen into truth and you shall know why you believe.[4]

Emerson is suggesting a fluid interconnection between the primacy of physical reality and emergent abstract knowledge about that reality. He is restoring an order of unfolding truth that resonates with real experience. In the process, he is turning on its head the tendency of modernity to deny the instinctual quality of reality in favor of an intellectual or abstract representation of sensuous experience. We have been programmed to distrust our gut instincts and direct engagement with nature, and to instead rely on our abstract rational thought processes to fulfill our intellectual potential and keep our "lower," instinctual self in check.

Today, we believe that having an instinctual connection to life impedes the unfolding of intellectual expression, but we are mistaken. There is no inherent contradiction. Our instincts are allies that complement our intellectual capacities, grounding our thinking in living nature. Our instincts help us tune into what Indigenous people refer to as "original instructions" for how to live on the planet—instructions similar to those embedded in the Scar Face story, where human beings learn how to develop alliances with other creatures and to properly respect them.

In short, the whole human being is not one who has *risen above* instinct, but one who has learned how to *include instinct* in the totality of their consciousness. But this is distinctly *not* what modernity believes. Instead, we fancy the idea that we have evolved beyond our animal nature, identifying with our rationality as the pinnacle of our development, and therefore all we are. This is why we justify suppressing our instinctual and emotional impulses. But what we do not realize is that, in doing so, we are also stifling our *spiritual potential*. Why? Because our spiritual growth is most likely to be activated when our *entire being* engages with what is beyond the (normal) human realm. This is hard to do of course, because the spiritual realms are beyond human understanding—which is why Indigenous peoples humbly pray for help in these realms. But Western society has done something far easier; we have chosen to accept as real only what we can comfortably consider within our narrow band of rational consciousness. And thus we dismiss whatever does not fit our preconceived notions.

This is why we generally counter one rational argument with another and become locked into polarized viewpoints. I am intending something different—to restore the definition of thinking to include multifaceted ways of perception that have been unnecessarily but systematically excluded from the full continuum of thought. This is why I venture into areas of story, dream, vision, and ceremony—areas where the rational mind is stretched.

My purpose is threefold: to expand the definition of thought and thereby expand what it means to be human—and then to reimagine the world we live in from that wider and more inclusive view. I take up the latter pursuit to the best of my ability in the final two parts of the book.

As I see it, the modern Western worldview has become unwittingly narrow. This occurred slowly over time as a result of certain choices that were made, often without awareness of the embedded assumptions governing the decisions. This is why it is essential to become aware of our tacit assumptions—because these assumptions, while they may open us to certain perceptions, blind us to far more. Below, I list what I consider to be the most common assumptions of modernity. I then proceed to deconstruct some of these assumptions for the purpose of getting to a deeper understanding of the human condition. I do so in a way that seeks to expose but otherwise avoid the common circular arguments (tautologies) that most modern belief systems rely upon.

Tacit Assumptions of Modernity Regarding Humanness

- Human beings are the (only) rational animal
- Instinct is a lesser way of knowing, relegated to the lower animals
- The human being is transcendent and superior to the rest of nature
- Only human beings are self-aware
- Human consciousness is separate from nature
- Reality is "out there"; what cannot be perceived with the five senses is not real
- Human beings are separate individuals; our thought processes are separate from each other's
- Human thinking and language are the products of higher (human) consciousness
- Human beings continually build upon the body of knowledge that came before, always improving ourselves throughout history
- Progress is (linear and) inevitable
- Human beings are not beholden to natural rhythms; we make our own timing

What Makes Humans Unique?

So what is it that makes humans unique? Is it the way we think—our supposed superior intelligence? Is it the arts and music we create, the ability to use tools, the ability to plan, the complex culture we live in, our ability to use language, our sense of humor? All of these attributes are unique to humans—or are they? Birds, crickets, and whales sing; other animals, including apes, monkeys, beaver, dolphins, and octopus, have been observed to plan ahead and use tools. Much, if not everything, on the planet speaks and/or communicates in some way; bees, ants, apes, and other animals have vastly complex societies; dolphins, otters, birds, dogs, and cats have an obvious sense of play, and so forth. As we watch animals more closely, the paragon virtues of humanity collapse one by one, and our assumed status as superior beings comes into question. Even our ability to think abstractly has been found in chimpanzees, who have demonstrated symbolic cognition abilities in learning both sign language and numbers. Given all the evidence, are we really as superior as we think we are?

It may be difficult to see, but our imagined superiority is protected from challenge by a process of circular reasoning. We assume that only humans are self-aware, and therefore the only species capable of considering the question of what makes us human. In this way, the question of what makes us human is conflated with notions about what makes us superior and transcendent from the rest of creation. Being self-aware and being human become indivisible; therefore, we find no need to look outside the human realm for signs of awareness.

If we were to carefully observe other animals, however, we would realize that our seemingly sound arguments are in the end specious—that self-awareness is not the province of humanity alone. When I take my dog for a walk and another dog approaches—of any shape, size, or color—my dog springs to attention. Clearly, she recognizes other dogs as akin to her. I am not saying that my dog maintains an abstract, taxonomic classification of canine in her brain, but she knows in her own way that other dogs are like her, and she invariably wants to interact with them. To me, this is a form of self-awareness. And when a bird calls to its companion to let it be known there are no predators in the area and that they have found a good food source, the calling bird is saying "C'mon over to me," or at least "C'mon over to where I am." This is another example of a form of self-awareness.[5] Animal awareness is quite different than human awareness; it tends to be hypersensitive to the immediate environment, for example— but that does not automatically make it less than human awareness. All animals are aware. All animals, as Darwin knew, have thoughts, feelings, and emotions.[6] I would also say that all animals have memories—which we prefer to call "instinct." But whatever we call it does not change the fact that animals can access a reservoir of experience unique to their particular species. This is why Indigenous people tend to say that animals have their own unique knowledge, and if we closely observe them, we can learn valuable information. I would agree. And animals observe us quite closely. If we could communicate adequately with our dogs, cats, cows, horses, and other familiar animals, I suspect they would have quite a lot to say about what makes us human.

Humans do have abilities that other animals (and other living things) apparently do not, including our capacity for philosophical self-reflexiveness (consciousness about consciousness), and the amazing array of technologies we make to manipulate our environment. But before we proclaim our particular mode of awareness as superior, we might want to look at the bigger picture. In particular, let us take a closer look at what we consider to be two of our species' most defining characteristics: rational thought and language. I believe that looking at the human species through these two lenses will help illuminate our question of what it means to be human.

Chapter 6

Rational Thought and Human Identity

How Did Man Come to Think of Himself as the Rational Animal?

Aristotle claimed that rational thought was the single most defining characteristic of human beings. His message is rarely disputed today. Yet, what we have come to know as "rationality" in modernity is only a pale approximation of its original meaning. Thus, while I acknowledge the importance of rational thought in human affairs, I believe it is time to reevaluate the equivalence of reason and humanness. Moreover, it is time to rediscover what rational thinking really is and how it originally flowered in human consciousness, so we can understand why it is declining today—and how we might reclaim its original essence.

The word "rational" originally came from "ratio," referring to harmony and proportion between things.[1] The ancient sense of rational implied an active engagement with nature and a recreation of that relationship in our thinking. This sense of involvement in the world is radically different than what rational means today; we now eschew involvement in favor of scientific objectivity. How did this difference come about? Let's explore, beginning with the ancient Greeks.

In ancient Greece, rational thought was just beginning to flower—unlike today, when it is declining. While it is true that rational thinking was seen, even back then, as the zenith of thought—it was also seen as the most *beautiful* form of thinking yet to arise. Greek rational thinking retained a sensual and instinctual relation to the world; there was a connection to beauty and the wholeness of nature. It was aesthetically pleasing to be a rational thinker back then—something hard to imagine today.

The Greek word for aesthetic perception—*aesthesis,* can be translated as "taking in," "breathing in," and "a gasp."[2] Aesthetic perception, for the Greeks, was "the speech of soul"; one's entire being was engaged.[3] To be human in ancient Greece was to be sensually immersed in the world, and the flowering of rationality did not sever this connection—at least not initially. Rational thought was an unfolding of a new way of seeing, but it was still connected with the emotional and intuitive aspects of perception. It emerged, like all perception emerged for the Greeks, from the Great Mystery—the "gasping" and "taking in" of the world. It did not and could not have distanced us from the world; it only served to connect us to life and all its wondrous beauty.

Our ancient way of thinking was thus life-affirming and vital—even as today's rationality is not. Rational thinking today is arguably the opposite of living thought; it is more of a disconnected scavenger feeding off the dead tissue of once-living knowledge. It is neither a creative nor an original way of thinking because it relies upon a body of already accepted (inert) knowledge. This kind of knowledge is not only dead; it is also partial and often misleading. And it has misled us about ancient Greece.

I do not mean to imply that our modern interpretation of Greece is a complete fabrication. There are certain aspects of ancient Greek thought—such as critical thinking and questioning of assumptions—that did at least partially promote the emergence of modern science. And we get to these shortly.

But our current interpretation of ancient Greece is, at best, half-true. We have selectively chosen only the aspects of Greek thought we deem useful—the ones we imagine to have been formative of modern thinking. We do this because our singular underlying assumption is that the modern view is more advanced. We are therefore predisposed to looking at what came before solely as a way of validating the way we think today.

In my opinion, the most troubling aspect of modernity is our belief of what it means to be human. This modern view celebrates rationality and individual free will as humanity's most essential attributes. This in itself might not be detrimental, but the peculiar kind of rationality that is chosen is. Why? Because it coldly moves us away from engagement with nature and promotes the idea that we are superior and transcendent from her. In doing so, it literally "rationalizes" our exploitation of the earth.

In short, we have retained only certain aspects of rationality, and discarded the heart and soul of its original intent. We once used rational thought to aesthetically appreciate nature; we now use it to clinically study it. Then, to top it all off, we justify our assumptions of our superiority by tracing our roots to an ancient Greek philosophy that never was the way we imagine it to be.

I am suggesting a different approach. Rather than justifying our assumptions, let's question them.

Critical Thinking and Questioning of One's Assumptions

There is little debate that Greek rational thought brought with it some very successful strategies for enhancing intellectual capacities—and that critical thinking and questioning of one's assumptions are chief among them. These capacities are essential to the overall palette of human behavior, and have become a particularly necessary aspect of formal education, a topic we will be discussing in detail later on. In brief, Socrates employed a method of questioning, founded on mental discipline, to penetrate superficial appearances and unveil hidden truths. Plato later formalized a lot of Socrates' techniques into a methodology he called the "dialectic" (or highest form of thought) and formed an academy that became the prototype of today's academic institutions.

Critical thinking is a particular mode of thinking in which a person continuously monitors the quality of his or her thought. Through this process, one gradually develops an increased capacity for self-correction, leading to improved accuracy and integral cohesiveness of thought. It requires rigorous discipline and mental attention—we might even say "mindfulness," to borrow a Buddhist term. Applicable to any subject or problem from the arts to the sciences, critical thinking assumes standards of excellence that have come to be known as the hallmark of higher education.

The questioning of assumptions was also integral to ancient Greek philosophy. In fact, Socrates questioned his own assumptions so radically that his only starting assumption was that everything he had ever thought or been taught was not true. Ever the contrarian, Socrates felt he had to continuously unlearn whatever he thought he knew and, in this ongoing fashion, he ceaselessly examined his own life. His life's work consisted of persistently probing and questioning others in a way that exposed the shallowness of their—often rigid—belief systems.

Clearly, the ancient Greek practice of critical thinking was formative to modern Western science. It was certainly foundational for the seventeenth-century (natural) philosopher René Descartes. Descartes, much like Socrates, adopted an attitude of radical doubt. He questioned everything, including many of the assumptions of his time and his own physical senses. Descartes, however, had a somewhat different motivation than Socrates. He sought to deconstruct reality in order to rebuild it with more reliable fundamental postulates or premises. These postulates lead to other postulates and imply the eventual accumulation of a body of knowledge, an essential ingredient of modern science.

Socrates, on the other hand, sought wisdom, not knowledge or the accumulation of knowledge. In fact, Socrates delighted in exposing the foibles of those who claimed to be erudite, and because of that, wise. For Socrates, wisdom came not from knowing—or thinking we know—but from freeing our minds from the grip of belief systems.[4]

The act of not relying upon accumulation of knowledge and instead trusting in the ability to receive knowledge as needed directly from Spirit is, to me, a key to thinking originally. It is why I began the book with a prayer pledging not to hold unto knowledge, but to offer it freely to those that will benefit. In so doing, I understand that whatever knowledge I offer is not really mine, but comes through me from an original source, a wellspring I trust to be eternal. I simply call that source Nature or Spirit; others have their own names. I believe that all of us, including myself, have access to this form of direct knowledge and that it presents itself differently in different ages, with each time, place, and culture creating a certain spirit of the age, or *Zeitgiest* of thought. I elaborate on this in Part Three (Return), when discussing the origin and circulation of paradigms.

It is my feeling that both Socrates and Descartes are miscast as historical figures that somehow rose above their cultural milieu—which is to say that we have effectively remade them from person into personage. In so doing, we unnaturally elevate and separate them from their times in order to validate our contemporary view, which sees history as both linear and progressive. We often do this unwittingly, because our paradigm demands it.

According to this view, Socrates was an iconoclastic figure, a sort of heroic mutant who changed the course of history by advocating rationality above all else. We see Descartes in a similar way, as the father of modern science.

In actuality, tradition and innovation coexist in every person. So, although it is true that Socrates advocated a *form* of rational thinking that eventually led to the separation of the individual from the tribe and from the natural world—at least in the West—this did not happen as cleanly or deliberately as we are led to believe. In the same way, Descartes was not singularly responsible for the "Cartesian worldview," a phrase too often bandied about. In fact, Descartes was intensely interested in what his contemporaries thought of him, and so conspicuously, that before publication of his famous treatise, *Meditations on First Philosophy,* the manuscript was circulated to a number of his contemporaries so that their comments and Descartes' responses could be included in the text as "Objections and Replies."[5]

Descartes is unfairly and often incorrectly cited by both mainstream science and those interested in spirituality for his role in separating mind from matter. Advocates for mainstream science look to Descartes as foundational; advocates for spirituality attack Descartes as a materialist. But, in his day, Hobbes attacked him as an *immaterialist.* In an objection written to the Second Meditation, Hobbes wrote, "We cannot conceive of thinking without a thinker. It seems to follow that a thinking thing is something corporeal."[6] Never mind for a moment that

Hobbes seemed to misunderstand that Descartes was being noncommittal at that point because of his posture of radical doubt, and that he readdressed the issue later on. Descartes was not simply interested in knowledge, but in *how that we know we know something.* The Meditations as a whole take us through a process of how one comes to know, and for this attempt, Descartes is to be commended, whether or not one agrees with his conclusions or even his method. I am personally more drawn to the way in which Krishnamurti and Bohm examine the thought process, and we will take this up in Part Three.

In my view, Descartes was a deep thinker, if not quite original in the way I am intending. He was also very much a product of his times. In his age, linear perspective was becoming the reigning way we saw reality; so it is only natural that he chose to emphasize cognition in a way that led to isolate thought from the body and the brain from other parts of the body. He did make the (dubious) statement that the "mind does not receive impressions from all parts of the body directly, but only from the brain."[7] He was also part of a shift in emphasis that isolated (and elevated) humans above other animals; and for this, I am less forgiving, as I would not put Descartes before the horse (if you will pardon the pun), or any other animal.

But Descartes is oversimplified when we attempt to say that he himself was responsible for the mind/matter split, the mechanization of science, or the silencing of spiritual thought. Descartes was a very religious man who believed that God was the author of his existence. His *Meditations,* lest we forget, were an attempt to prove the existence of God and the soul.

Moreover, Descartes strongly believed in direct knowing, which he alternatively referred to as either thought that he has "received from God," the "light of nature," or "clear and distinct perception."[8] All these expressions imply something similar to what we mean today by "intuition" or direct perception of truth, but curiously, intuition meant something else in Descartes' time. The original sense of intuition was: "to contemplate" or "look at," from the Latin stem *intuitiō.* Descartes' *Meditations on First Philosophy*—the entire treatise—could be considered an intuition in the original sense of the word.

I find it significant how much the word intuition has shifted meaning. In Descartes' time, there was not an opposition between intuition and reasoning. The two coexisted together, for intuition was still considered reasoning in the form of sacred contemplation or meditation. Today, however, intuition means a direct perception apart from reasoning. The consequence of this belief is that rational and intuitive thought have become polarized.

The modern bifurcation of the rational and intuitive implies that we are not thinking as well (or not at all) unless we are using reason. Accordingly, the complete spectrum of consciousness is divided into activities deemed separate and distinct: sensation, feeling, thought, and intuition. Once thought was separated from the physical, the emotional, and the intuitive, it became solely associated with the mental/rational. Human beings, by default, have been largely relegated to the same.

The entire society, men and women alike, have basically accepted the equating of thought with the mental/rational aspect, and identified with it, with barely a whimper of protest. But the relegation of serious thought to the mental/rational realm has had devastating effects upon all people, women and men, and especially upon women and nature. Why? Because the cut and dried type of thinking we associate with the mental/rational—"time is money," "business is business"—can be brutally uncaring, and has been seized upon by men in power to justify both the suppression of women and nature. It is not an exaggeration to say that the language of modern rationality is biased toward the masculine values of analytical separation. Division implies control, as in "divide and conquer."

Women and nature, both of which were seen as unpredictable and needing control, were devalued. Indigenous peoples suffered a similar plight, and this too, is not surprising because the word "nature" comes from the same root as "native," "natal," and "nativity."[9] Women have always been linked to nature because they are the life-givers who give birth. Indigenous peoples were often lumped in with women in the sense that their traditional knowledge, along with women's "old wives tales," were (and still are by many) considered obsolete in the wake of linear and progressive rationalism.

The only problem is that the whole argument is false. Modern rationality is not improving; it is degrading. And it is based on a faulty premise that our forebears (the ancient Greeks) were the same as we are today, when in fact they were far less cerebral than we make them out to be. Rationality, in the time of the ancient Greeks, was a vital, organic philosophy of life. The Greeks valued thinking, yes, but thinking that was elegant and appealing to their soul—which is why Socrates spoke of thought as potentially more enchanting than the beauty of form.

Modern rationality is a lesser form of rationality than that of our forebears. It is thinking that has been cut off from the senses and from all of nature and its concomitant sources of inspiration—from the birds, the trees, and the myriad of ways that creation speaks to and through us.

In ancient Greece, there was not a dichotomy between science and the arts. But now a wedge has been inserted between two aspects of thought that ultimately come from the same source—and this has created a false dichotomy. The way we think about thinking has become increasingly narrow and limited.

Descartes, in his own way, embraced a wider concept of thinking than we would typically grant today. When Descartes made his famous pronouncement: "I think, therefore, I am," he was not simply equating humanness with rationality. He addressed the question head on in the Second Meditation. "What is a man?" he asks. "Shall I say a rational animal? Certainly not."[10] He then goes on to explain other aspects of being human, including having a face, hands, arms, requiring nourishment, and so forth. This is similar to Aristotle, who did not say that man equals the rational animal; he said something more along the lines of humans having the capacity for rationality (or "deliberative imagination") in addition to other capacities shared with the plant and animal kingdoms.

We are not human because we think rationally or in any one particular way. We are human because of the combination of ways we express ourselves. Many years ago, in the dialogues, Donna House said the Diné people traditionally believe that: "We think *and sing* our reality into being." This is something I immediately resonated with. It was also something that Donna's Diné "sister," Pat McCabe (Woman Stands Shining) used to regularly do during the dialogues: break out in song.

The act of creation may require the vibration of song, and perhaps even some laughter and joy, which reminds me of another thing that once happened at the dialogues, something that in a playful way spoke of the opposite of creation. The Nobel Laureate physicist, Brian Josephson, shared a story of a not very well-known incident in Descartes's life. As Josephson tells it, "He was having coffee in a French restaurant, and the waiter said to Descartes, 'Would you like some cream in your coffee, Mr. Descartes?' He said, 'I think not.' And, suddenly, poof, he disappeared."

How Did Rational Thought Become Disconnected from Life?

At the risk of making Aristotle into a personage, we must acknowledge his role in accelerating this transition. A key difference between Aristotle and Socrates (and/or Plato) is that Aristotle believed that knowledge was derived from experience, whereas Socrates and Plato believed that knowledge existed a priori (before experience). In other words, Socrates and Plato believed that all knowledge already exists within us, and that humans must discover it in the archaic (original) sense of to reveal or make known. Consequently, Socrates did not teach by lecturing; rather, he prompted students to think for themselves by asking questions and challenging his students to discover hidden knowledge and wisdom. Plato largely followed in the way of his teacher, believing we are born with knowledge that we must recollect (re-collect) rather than learn anew.

Aristotle, however, believed that the knowledge gained by experience could be taught to others. This belief is the precursor of the current Western educational model, which rests on the assumption that education constitutes the transmittal of a body of knowledge from teacher to student. This initially was more of a master-apprentice relationship in which the teacher physically demonstrated the tools of the trade or art form. But with the creation of formal education, increasingly abstract information was transmitted from teacher to student, until the accumulation of knowledge was no longer associated with direct experience at all. Neither students nor teachers necessarily possessed knowledge obtained through direct life experience. We advanced beyond our dependence upon nature to reveal living knowledge to us in the moment; instead, we increasingly relied upon the stored knowledge of humanity. This is called progress, even as it distances us from nature, the original teacher.

What Is the Role of Technology in Human Development?

Technology has been with us for a very long time, and it has been probably always a mixed blessing. We think of the earliest technologies as less invasive and more aligned with the living wisdom of the earth, but this can be misleading; for example, the advent of agriculture is in some ways one of the most dramatic changes we have ever made. We went from being in and of the earth—participating in the renewal of cycles as they manifested in the movements of the sun, the rain, and snow; the migrations of animals; the return of berries and seeds, fruits and nuts—to planting our own food and forming villages around that central agricultural hub. But we still had to retain our connection to the seasons of change if we were to be successful growers.

It is not until the post-Renaissance period that technologies designed to outwit Mother Nature were aggressively developed. The European Renaissance began with high aspirations to recover the wisdom that had existed in classical Greece. I am a great fan of Renaissance art, architecture, and mathematics. However, after the initial creative impulses of the era, something else arose.

Specifically, technological innovations emerged to take advantage of creative inspiration. This process, which still shapes our lives today, follows a certain pattern. Each subsequent innovation is a further abstraction from the original inspiration. We can see this in the evolution of communication as it has moved from live speech to written text, to the printing press, and then the bits and bytes of the computer.

Technological innovations are often quite useful, even as they are sometimes made out to be greater than they actually are, particularly in a society where innovations are often the result of the planned obsolescence of a previous product. But even if the innovation is substantive, it still represents a degradation of rational thought. Here's why: Creative genius arises from an unlimited, original source; it comes from nature and moves through us. But technological innovation arises from a more restricted source; instead of drawing upon all of nature, it incrementally builds upon other human inventions. In this way, technological innovation represents a narrowing of consciousness; it turns away from the unlimited font of nature to tinker with the end products.

This focus on end products—rather than origins—leads to some misperceptions, such as a tendency to see human beings as the primary source of inspiration, or the idea that we can improve upon nature. We would be better to follow the aforementioned guideline of Buckminster Fuller, who knew that no invention could be achieved if Universe would not allow it—in other words, if it did not already exist in "universal mind." Instead, we have come to believe otherwise—that we are the one species that can rise above and, hence, improve upon nature. Once we imagined ourselves—our own ego—to be the source of inspiration, we no longer needed to align with nature, or so we thought.

It is in this way that the West began a bold and dangerous venture, setting a precedent for how human beings were to live—not just apart from nature, but apart from all the perennial wisdom that came before. The key difference is that previous civilizations believed that the human must *follow* nature. The Tao Te Ching, for example, states: "Man follows earth. Earth follows heaven."[11] But, at some point in the West, man stopped following earth in an attempt to skip a step in the natural order. Man still sought the divine, but now *directly*, with our own intellect, our own cleverness. Rational thought became elevated as high discourse—a shortcut to the Gods. The Earth, no longer seen as our teacher, became associated with Satan or uncleanliness. (Dirt became dirty and needed to be sanitized).

The intellect, courtesy of Descartes and others, became increasingly located in the brain. The brain became considered the organ of thought to the detriment of the heart, soul, and body, which were more associated with the Earth. As a consequence, human beings, particularly in the West, became both ungrounded and headstrong.

The Western fascination with head knowledge reminds me of something that happened to me over thirty years ago that has haunted me ever since. I was driving on a freeway in the middle of Pennsylvania when I witnessed an accident. It happened right in front of me. Two cars, one a convertible, side-swiped each other. The convertible turned over and skidded to a halt. I had to swerve to miss them but was one of the first people there to attempt a rescue. The two drivers turned out to be brothers. One was banged up but relatively unscathed; he must have been wearing a seat belt. He was able to get out of the car with our assistance. His brother, however, was not so fortunate. He remained trapped underneath the vehicle with his head bleeding profusely, and had probably dragged his head on the pavement for quite some time. The one that we rescued became hysterical as soon as he saw his brother. He kept yelling at him over and over again in fear, which soon turned to despair. But his brother, turned upside down and unable to see how much his own head was bleeding, kept calmly trying to assure him he was all right. I thought about organizing the group to turn over the car right then, as that may have been our only chance to save him. But that is not what we did. Instead, we did what we were supposed to do—call for help. Calling for help meant going for help, because this was long before the era of cell phones. I volunteered, and drove to the next exit and called for an ambulance. After that, I do not know what happened, and I could not have gotten back to the scene if I wanted to because I would have been hopelessly stuck in traffic.

Sometimes the events of real life seem to signify something else, which is why I tell this story. I came to see these two brothers as a metaphor for the degradation and potential death of Western civilization. One brother represents the rational mind, and the other the intuitive/emotional. The rational mind thinks it has everything under control, even as it is bleeding to death. The emotional/intuitive mind is repressed and then becomes hysterical. Both are out of balance. We need to awaken to the clear and present dangers we face—nuclear proliferation, global warming, lack of clean water—before it is too late. And we have to respond to these dangers with our whole being, not just our rational mind. We have to move in balance; we cannot just scream hysterically. But we also can no longer rationalize our situation, believing that we have things under control. We do not.

One of the things we cannot control with our minds is nuclear power. Paula Gunn Allen used to say that what is sacred is "powerful and a little scary," like the "yellow cake" (uranium) that is so plentiful in the land of her childhood (outside Laguna Pueblo, NM). The yellow cake should not be messed with, she learned, which meant it should stay underground. Things that are sacred need to be treated with respect. Long ago, I am sure we understood this. Today, I am not so sure.

We must learn to respect the yellow cake, and not try to control it. It is the desire to control that prompted us to mine uranium, and to break apart the atom. We messed around and are still messing around with what ought to be considered sacred.

It was not supposed to be this way. As modern science developed, there was a form of "gentleman's agreement' (or other form of unspoken truce), between religious leaders and scientists. Religious leaders agreed to eschew meddling in the world of material forces so long as science did not intrude upon their jurisdiction in the transcendent realms of the sacred and spiritual.

This schism between science and religion has born mixed results. It gave science the freedom to develop great inventions and new technologies, but it also gave science a license to turn our living and sensuous universe into an experimental playground, a vulnerable and largely unprotected laboratory where they could mechanistically apply rational thought unguided by a moral or ethical compass.

Today, science is doing some amazing things, but many of them are controversial. One wonder drug—ATryn, is an anticoagulant that has successfully been used to prevent life-threatening blood clots. It is also the world's first transgenic animal drug, produced from goats that have been genetically modified by injecting human genes (a therapeutic compound antithrombin) directly into fertilized goat eggs and then implanting the eggs into the womb of the female goats. When the transgenic female goats lactated, they produced a milk containing the vital human medicine.

But that is not all.

The same or similar microinjection process has produced transgenic "frankenfish" like the Atlantic salmon that grow at enormous rates aided by genes from a Chinook salmon and an ocean pout, an eel-like fish that lives in frigid water. This way, the salmon can grow all year round instead of only in the summer as nature originally planned. There is also an entire industry of ornamental fish that can glow in various colors because of the injection of foreign DNA, including a popular Glofish (a goldfish that glows). In China, there is an enormous industry of mutant mice. Thousands of mutant mice—about 50,000 in Shanghai's Fudan University alone, are being bred assembly-line style. They are being genetically altered to grow everything from tusks to skin tumors to male-pattern baldness. Some mice cannot feel pain; others can only turn in one direction. Cockroaches, flies, and beetles are being remote controlled and may soon (if not already) become a cyborg insect army or mini-drone spies.

It gets weirder and weirder as the list goes on. My source for the above is the fascinating, if scary, book by Emily Anthes, called *Frankenstein's Cat: Cuddling up to Biotech's Brave New Beasts*.[12] Anthes does her best to present these cases without passing ethical judgment, even as she claims to be an animal lover. She points out the many incidents where these innovations save human lives, and that not all of this is new and futuristic (we have been breeding dogs for centuries). Not every experiment is to the detriment of the animals; some have helped them, such as the outfitting of sandhill cranes with synthetic prosthesis after they have been hit by wayward golf balls and cars, accidents that normally would have resulted in them being euthanized.

She makes one observation, however, that cuts to the chase. She simply says, "Making animals more useful also makes them more likely to be used."[13] My own complaint with biotechnology goes beyond the using and harming of research animals, although this was my first reaction, and why I refused to dissect animals in college. It goes beyond the fact that we are moving around the parts of nature for specific human-centered purposes, playing God. It is not that biotechnology is unethical, even as I believe it is. It's that we do not know what we are doing or what the ramifications and consequences of our actions will be. To rearrange the parts of nature is to invite a multitude of disastrous consequences that we could not possibly account for.

It would be better, in my view, to limit our innovations to mimicking what already exists in nature. The natural world is chock full of amazing examples of adaptive innovation that we know to be harmonious with other creatures. We could study these naturally occurring examples from now until forever, and it would not necessarily slow down our innovative efforts on behalf of humanity. The counterpart to Anthes book is Janine Benyus' classic: *Biomimicry*.[14] Benyus details an array of fascinating and creative lessons we can learn from the diversity and talent of the natural world. At the end of her book, she summarizes ten lessons we can learn from nature, including using waste as a resource; diversify and cooperate, optimize rather than maximize, and so forth.[15] The whole list is not important for our purposes; the point I wish to make is simply this. Technology can be grounded in nature, and if we do so, there is a better chance that the technology will work, will do no harm, and will be sustainable.

Human Identity and Technology: Preserving the Integrity of the Whole

It is imperative that we come to grips with technology—what it is and how it affects our life and that of the planet—because if we do not define technology, it will define us. And it already has to a large degree. Why else would we define our progress almost exclusively in terms of what technological advances we make?

We must remember that technological innovation is not true scientific discovery, but rather the application of the known for specific and limited purposes. However enticing the products it produces, technology is always based in "partial knowledge." The partial knowledge our devices give us is being confused with the original knowledge of nature that is whole and complete. We have become dependent upon "googling" knowledge to the exclusion of opening ourselves up to creation. We are all capable of receiving knowledge directly. It is our birthright.

The allure of technological development has had an unfortunate and little recognized consequence, which is that we increasingly identify with technology as an appendage of ourselves. We are not our machines. But as long as we let our machines define us, we will become increasingly machine-like in our thinking. How far do we want to go with this? With the introduction of "Google-glass," we now have the opportunity to wear glasses that are computers. Is becoming a sci-fi cyborg what we really want to do?

My biggest concern is the effect that technology will have on our thinking, and specifically on our ability to see the bigger picture. I am interested in thinking that promotes and preserves the integrity of the whole, and it is in this context that I evaluate technology. I am not a modern Luddite; I use many modern conveniences myself. I drive automobiles, fly on airplanes, type on computers, and type on computers on airplanes. I do have a fondness for technologies that connect me to nature. Some of them are very old and have not changed much over time, such as water catchment for irrigation and sustainable gardening purposes (now called permaculture). There is nothing wrong, in my estimation, with using the latest, most efficient solar panels or driving the latest and most efficient hybrid vehicle. But the most important technology of all—our mind—is both ancient and new at the same time.

Feminine and Masculine/Right and Left Brain

It is no secret that the past two millennia of patriarchy have seen the privileging of certain ways of knowing, those considered more male (exacting, logical, abstract) at the expense of equally powerful, female-centered ways of knowing—ways more intuitive, connected, and whole. This hegemony not only led to (and helped justify) the suppression of women, but it has helped rationalize our exploitation of the Earth and its resources. The pendulum is now swinging back the other way, of course; and this time, according to Mayan and other prophecies, we are coming into an age of balance rather than dominance by one side or another.

For the time being, however, we can speak of two distinct ways of seeing: one that seeks the greater whole, and the other that divides the world up into conceptual parts. These two ways of seeing may seem to be principally feminine and masculine, but they are actually universal, contained within every human being, female or male. They are different, but complementary; together, they can make us whole and balanced people, but we often experience them as contradictory, dividing the self against itself.

These distinct ways of seeing are reflected in the human brain, which itself is divided into two hemispheres. Curiously, while the two hemispheres can share information, they also have a seemingly contradictory need to keep their different ways of perceiving separate. This tendency toward mutual inhibition sounds a lot like competition, but it could also be looked at as *cooperation that requires difference*, not more of the same.[16]

Perhaps it is no coincidence that what we think of as primarily left brain functioning seems to carry greater weight in modern society. It appears that the left brain, which requires the isolation and separation of phenomena into parts in order to comprehend things, has succeeded in dominating the right brain, and this preference for analytical thinking has then become the dominant force in the greater society. The situation is, in actuality, more complicated than this. While the left brain often operates as if it can function on its own, it cannot; it very much needs the right brain. Studies of patients with damage to one or the other hemisphere have convincingly shown this. Moreover, they have demonstrated the converse: it is the right side of the brain, not the left that can better function on its own. This is because the right brain is "the primary mediator of experience," according to psychiatrist Ian McGilchrist, "from which the conceptualized, re-presented world of the left hemisphere derives, and upon which it depends."[17]

The right brain of course also needs the left, just not in the same primary capacity. The right brain needs the left brain to do its bidding—to get things manifested in three-dimensional reality. The right brain intuitively grasps the whole of a situation, but intuition alone is incapable of action. Conversely, action without intuitive knowing is blind. Both hemispheres, and both ways of knowing, need each other, and we need to use each of their gifts appropriately. But this is not what tends to happen presently, particularly within Western cultures. The left brain, or executive function, tends to be dominant, mirroring the larger imbalance in society toward male dominance.

The problem is not only that the left brain is directing all the action—but that it is doing so without first seeking the guidance of the right brain, which has the capacity to see the whole. As a consequence, many of our thoughts and actions are "half-brained" and ungrounded; they go off on a tangent—unrelated to the larger situation—and create unanticipated, often negative consequences. Because these tendencies are associated with gender, it is possible that the deficit can be ameliorated when the balance of male and female is more equal in the halls of leadership. But it is not just gender relations that need rebalancing. We need to allow the creative, intuitive aspect (associated with the feminine) to come forward in all of us—so that our thoughts and actions come from a deeper place of wholeness and are therefore more original, wise, and effective.

Awakening from the Sleep of Modernity: What Are the Implications for What It Means to Be Human?

As modern humans, we are increasingly absorbed in thought—rational thought—and caught between two worlds: the outer and the inner. And, increasingly, the noise in our head is being shaped and mediated by our own technology, such that our human experience is becoming more and more detached and self-referential.

Many of us are entrapped in the shell of our minds, and this follows us wherever we go. We wake up each morning, and rather than taking a moment of appreciation for the miracle that is life, we snap to mental attention. We make an immediate inventory of our egoic self—who we are, what we are, and what we are going to do today. We remain in the shell of our thoughts even as we leave the secondary shell of our enclosed houses, sipping our morning coffee as we enter the tertiary shell of our automobiles, backing out of our driveways, barely noticing whether the sun is shining or the birds are chirping. Bucky Fuller once said that if an alien being were standing right in front of us, we would not notice because we are too stuck in our way of seeing. This is probably true, but more importantly, we do not notice what is natural to our world any more than what is alien. We have done our best to insulate ourselves from the sensuous life-world all around us; we move from place to place within the shell of our retracted consciousness, literally lost in thought.

The sad fact is that we moderns spend so much time thinking that we ignore life's beauty and vibrancy. And, even sadder, our thinking processes have wrapped us so much in knots that we have come to believe that our thoughts are all there is, and that the rest of the world is dead.

Let me put it starkly. The sum total of the modern paradigm is a belief that human beings are separate, superior, rational, and alone in our consciousness on a dead planet, without purpose.[18] We have all inherited a worldview that is bereft of soul and stripped of life, cut off from our spiritual roots.

This would be exceedingly depressing if it were true. But, thankfully, it is not. The truth, as I have come to see it, is that we humans are not superior; we only think we are. We are just different. The planet is not "dead" or a mere machine; it is all alive. And we are never separate or alone in our consciousness, no matter what we might imagine. Some part of us has always known this. The very word "consciousness" originally meant *knowing with*, from the Latin *conscientia*, or shared knowledge. The fact of consciousness is a radical expression of interconnectedness; everything that exists contributes in some way to the consciousness of the whole.

It is not that rationality—as we have come to define it in the modern word—is unimportant. It is important. But it is only one expression of our humanness, one part of our full potential. It does not, should not, define humanity. We are not only mental beings; we are whole human beings with physical, emotional, intuitive, and spiritual aspects. We know this, and yet most of us still favor rational thought, even if unconsciously. We do this when we compartmentalize our lives, itself an act of rational thinking. Our spirituality, for instance, may be relegated to church, prayer, mantra, or meditation; our emotional life may be limited to interaction with our spouse and closest of friends; and our intuition, for many of us, is completely suppressed or marginalized. We, both women and men, but especially men, are trained to follow our head rather than our heart, to distrust our gut instincts and intuition. All of this reinforces and artificially inflates the importance of rational thinking.

In actuality, our intellectual capacities are useless unless they are connected to an inspirational source. When we exercise Herculean efforts to control the world with our intellect alone, it just does not work. We may think we are "masters of the universe," as the Wall Street sharks used to say, but we are in for a fall later on. Even when we accomplish what we set out to do, it (intellectual savvy) does not equate to long-term happiness or satisfaction. We keep asking ourselves, where did we go wrong?

How can we recover the sense of interconnectedness we once enjoyed? To do so, we must go beneath and beyond rationality to uncover and rethink the ways we originally sought to understand our world and forge meaning in our lives. The original way that humanity created meaning was through language, which is why we must explore the origins of language if we are to do justice to our question of what it means to be human. Like rationality, language seems to define us, to set us apart from the rest of creation. But is this really so? What are the deep roots of language? And how does language shape our idea of what it means to be human? This is our next subject of inquiry.

Chapter 7

Rethinking Language

Is Human Communication Really the Most Advanced Form of Language?

The mainstream view holds that human language is evidence of our innate superiority. We are said to be possessors of a "higher" language, one that is an evolutionary outgrowth of brain development, specifically our frontal lobes. At first glance, there seems good reason to believe this. Human beings do, in fact, have larger frontal lobes than other animals relative to their size, and human communication is unquestionably sophisticated and becoming more so all the time. As a species, we speak in many different tongues, including abstract languages, like mathematics and computer languages.

Upon closer examination, however, our superiority is not so evident because it is based on certain assumptions we rarely consider. One assumption is that brain size equals intelligence; another is that intelligence produces language and that the temporal lobe (in particular, the left temporal lobe) is responsible for language development and communication.

The assumption that brain size equals intelligence is easy to overturn even within the human species. There have been very intelligent people with relatively small brains, and large-brained people who do not appear particularly bright.[1] The association between frontal lobes and language development is flawed for a different reason—because frontal lobes have different functions in different species.

Our flawed perception begins when we go outside humankind to look for signs of intelligence and language, for we tend to look only for the specific kinds of intelligence and language skills that we already are familiar with in ourselves—in humans. But there is tremendous diversity of intelligence in nature. Consider cetaceans. Dolphins, whales, and porpoises all have brain-size-to-body ratios similar to that of humans, and all have large cerebral cortexes and large frontal lobes, believed to be associated with abstract thought. But the cetacean brain has evolved in ways very different from the human brain—differences obviously related to the radically different environments each species has to negotiate. For example, dolphins have ten times our capacity for acoustical imaging, undoubtedly related to their use of echo-location in water (which is a superior medium for transmission of sound).[2] Humans, on the other hand, have comparably superior visual abilities, which serve us well upon land.

What significance does this have for our understanding of language? In humans, abstract thought is a projection of visual images, which we then try to express by moving air through a passage to our vocal cords: that is, through speech. This indicates that our primary means of communication is significantly different from the way we think. But cetaceans think and perceive in auditory images, and they apparently express themselves in the same way: through auditory images! This gives cetaceans a distinct advantage over us, according to Marable, in that "their primary sense is the same as their primary means of communication."[3] We do not really know that cetaceans regularly communicate in this manner, but we know that they can—that it is possible for cetaceans to project an auditory image that is identical to the sonar image they receive from scanning their environment. This is equivalent to human beings sending and receiving holographic images to one another.[4]

This brings up another aspect to language that is rarely considered. It seems that most, if not all, animals (other than human) as well as plants and trees, are adept at communicating their intent without necessarily vocalizing it. If one tree in a forest is chopped into, for example, the other trees instantly know about it. We explain this away as an instinctual "chemical transmission," but it may be an entirely different phenomena. I suspect it is a form of interconnected consciousness that we have trouble understanding, despite the fact that such interconnectedness abounds in nature. One out-of-the-box way of looking at this was provocatively expressed by Little Bear when he began one of the earlier dialogues by saying that "language is a sign of primitiveness." In other words, language (oral and written) is needed only until a more direct means of communication is developed.

As playful as Little Bear's statement is, I suspect something else is at work, and that he knows it. I believe we once routinely had the ability to communicate with all of nature without words, and that it was the advent and overuse of rational thinking that eventually suppressed our once robust abilities. I have reason to suspect this because, in cultures that are not as fixated upon rational and abstract thought—–including but not only Indigenous cultures—there is greater evidence of interconnected consciousness. Conversely, in cultures such as ours that strongly emphasize the development of abstract thought, those few individuals who can still perceive interconnected consciousness are considered extraordinary, that is, they are considered "psychic" and exhibit "paranormal" abilities. It is not so much that they have a special talent, but that they have not forgotten. For the majority of nature, interconnected consciousness is the norm; evidence of it abounds if we only look for it.

The point I am making is that intelligence and the capacity for language ought not to be evaluated by predetermined means that are specific to species or a particular subset of a culture. We assume that certain kinds of language and intelligence are more advanced, such as the vocalization of abstract thought through words or the expression of words in writing. Animals (and even other humans) who operate differently are prejudged as inferior without due consideration of the diversity of means of expressing linguistic intelligence.

Our blind spot extends even beyond this. When other animals perform in specifically human ways, such as when parrots learn to speak English, we still ignore or dismiss this. Birds could not be intelligent! They have only "bird-brains," we say pejoratively; a parrot can only repeat or "parrot" back to us human words. We are surprised to learn from biologist

Rupert Sheldrake's research that there are numerous documented cases of parrots composing contextually relevant original sentences.[5] How come we never noticed this before? Could it be that we did not notice it because we considered it impossible? I suspect so; I would venture to say that the most frequent reason "anomalous" results are dismissed, or simply go unreported, is because they are outside the paradigm of the researcher and therefore they either do not see it or do not report it. One such anomalous report, and a rather amazing one at that, is of dolphins who have learned to speak rudimentary English through their blowholes! John Lilly taught his dolphin companions to do this.[6] I have heard the audio, and this dolphin speech is perfectly understandable. I am still waiting for humans to learn to speak dolphin languages, rudimentary or otherwise.

None of this is convincing to linguists, who deem human language to be the most highly developed and sophisticated form of communication. This is their position, and they are sticking to it. While virtually all linguists and scientists recognize that other animals are able to communicate, it is human beings alone, they say, who possess the physical hard wiring that enables the complex thinking processes and symbolic cognition necessary for (human) language. This view, seemingly logical, contains a hidden bias; again, it is circular reasoning to say that only humans possess the physical wiring for human language. Would it not be just as reasonable to propose that all creatures possess the inherent characteristics that enable each species to develop whatever form of language best enables them to interact/communicate most effectively with their particular environment?

Where Does Language Come From?

There is an ancient idea that language emerges from the land itself—that language coarises with our experience of place. When I was in Costa Rica, immersed in the medium of sound, I felt this; it was no longer just an idea. In the rainforest, I felt vibration pulsing in and through me. It was not only the immense variety of bird calls, or the chattering of monkeys, the powerful rushing of rivers, the scent of rain, or the wind snaking through the trees—much as all of this left me awestruck. It was not any one thing, but the confluence of many. It was the merging of the sounds and sights all together that suddenly broke through my awareness. At that moment, I realized, yes! Sounds can beget language, and vibrations can crystallize into words. I am a man, and I have never given birth, but I know what it is to be pregnant, at least pregnant with language.

This idea of language emerging from the land may seem far-fetched to some, but recent research conducted by the Terralingua Foundation supports it. Their studies find an astonishing correlation between biodiversity and linguistic diversity. Specifically, ten of the twelve countries possessing the greatest biodiversity are also among the twenty-five with the greatest linguistic diversity.[7] The Terralingua research points to a profound relationship between the ecology of a place and the languages it gives rise to. It seems that when we cut down a rainforest, we are not only diminishing biodiversity; we are also severing language, both human and nonhuman, from its source.

If the evidence of Terralingua is true, and it seems it is, then it follows that the origin of human language may be traceable to the linguistic soundings of a place as it mixes with human consciousness. The ramifications of this are astonishing. It means that language is not a product of the mind alone, but the result of relationship.

Are Language, Land, and Worldview Interconnected?

The idea that there is a relationship between language and land is not new to Native Americans. It is what Chickasaw elder Sakej Henderson calls the *langscape*. If language comes from the land—first in soundings, later becoming words—this would explain a lot about the diversity of human language. It would explain, for instance, why particular languages seem so exquisitely appropriate for their specific environments, as was noted by linguists Benjamin Whorf and Edward Sapir in the middle of the twentieth century. The majority of modern linguists dismissed this insight, siding with Chomsky, who proposed that all languages are equally adept at describing the world.[8]

But outside of academe, the insight had resonance; the idea that Eskimos have more words for "snow" took root in the popular imagination, and not without cause. Yupik and Inuit peoples of Alaska do have different words for snow because they need to distinguish between different qualities of experience. To them, there is a vast difference between falling snow and snow on the ground, hard packed snow and slushy snow, snow softly falling or wind-driven flying snow, and so forth.[9] They also have more words for other things important in their environment, such as the movements and activities of caribou, whales, walrus, and polar bears, and anything else integral to the unfolding processes of life peculiar to their lands. In the same way, Pacific Northwest tribes have more words to describe the movements and processes of salmon and salmon fishery; Anishanaabe people have more words pertinent to cultivating wild rice; southwest Pueblo tribes have more words related to the planting of corn; and California "Valley girls" have more words relevant to contemporary style and fashion.

This does not mean that our language *determines* the way we see the world, although linguists used that straw-man argument as a way to attack Whorf and the Whorf-Sapir hypotheses. Neither Whorf nor Sapir ever said that.[10] Whorf did say that "every language is a vast pattern system different from others," and that "the forms of a person's thoughts are controlled by inexorable laws of pattern of which he is unconscious"—but he did not isolate language as the sole cause of thought or worldview.[11] Whorf took a much broader view, one that looked for the complex and reciprocal interaction of language, thought, and reality—a view that is in consanguinity with the spirit of this book. He did so because he was an original, transdisciplinary thinker, unlike his critics, who were linguistic specialists.

In short, Whorf saw nature and language as akin. Nature is composed of patterns that form wholes and so is language. But individual languages selectively emphasize only certain patterned relations within the structures of their particular grammars. They take the entire flow of existence and recreate it within particular sentence structures. This predisposes the speaker to make certain types of observations and judgments, which in turn influence the speaker to think and see his or her environment in a particular way. This can lead to habitual ways of thinking and, hence, divergent worldviews. But language does not force us to see things one way. It is more, as Little Bear says, that language "leads us down a garden path." And that path is a two-way street. It is not only our grammar but also our environment that interacts with our consciousness, shaping both the way we think and the way we compose language. In this way, language, thought, and environment all work together to create our perception of reality.

The idea that humanity's defining characteristics—language and thought—originate from interaction with nature, runs counter to the tenets of modernity. In the modern West, we believe that thought and language enable humans to dissect, describe, and manipulate reality—and that is what makes us superior. But if language actually emerges from our interaction with our environment, and if our consciousness is not superior so much as coevolving with all consciousness, this would say something very different: it would say that humanity is deeply interconnected with the rest of creation. It might even explain Indigenous stories that speak of a time when all the Earth's creatures spoke the same language, and communication was not yet divided by species or cultures.

> *In the very first times …*
> *both people and animals lived on the Earth*
> *But there was no difference between them.*
> *A person could become an animal*
> *And an animal could become a human being.*
> *There were wolves, bears, and foxes*
> *but as soon as they turned into humans*
> *they were all the same.*
> *They may have had different habits*
> *but all spoke the same tongue*
> *lived in the same kind of house*
> *and spoke and hunted in the same way.*
> *That is the way they lived*
> *here on Earth*

in the very earliest times——
times that no one can understand now.
That was the time when magic words were made.
A word spoken by chance
would suddenly
become powerful
and what people wanted to happen
could happen
and nobody could explain
how it was.[12]

A similar kind of story, of one language spoken by all, is memorialized in the biblical telling of the Tower of Babel—the only difference being that common language is confined to humans. And there are many other reasons to believe that our oldest languages came directly from the land. Native American and aboriginal languages, as well as Sanskrit, Hebrew, Arabic, and other ancient languages, are sometimes called sacred languages because the soundings are known to evoke the same vibrational frequency as the movement of energies in nature. This is why singing and chanting are so important in Hebrew, Arabic, and Sanskrit prayers, as well as in many Indigenous languages. This form of prayer is known to produce changes in existing vibrational patterns. Such a prayer is not a plea for an action at some future time, but an action in and of itself.

What Has Been Lost as Our Language Has Evolved?

More recently developed languages, such as English and other Indo-European languages, may have had similar sacred origins, but they now operate quite differently. With the possible exception of vowels, their sounds no longer evoke the movement of energy in nature; this connection, once so deep, has been severed. Instead, the sounds are almost incidental, mere vehicles for conveying a representation of the physical reality, not replicating or participating with it directly.

This alteration in the function and form of languages has had, over time, the net effect of removing human consciousness from our environment, in complete contrast with our earliest languages, which served to connect us. As David Abram artfully unveiled in *The Spell of the Sensuous,* it was the transition from spoken to written language that largely muffled the sounds of nature's language, once heard by all. The written alphabet (drawn from the original sacred sounds of the Greek "aleph-beth") is one step removed, providing only a pale reflection of the original power of the spoken word.[13] To recapture the immediacy of the spoken word, the reader of this or any other book must call upon their archetypal imagination, but it is still there, waiting to be reawakened—a deep-seated, original memory of nature speaking to and through us.

The transition from a reliance on the spoken word to written language has been pervasive, and its effects profound. It has marginalized (if not eradicated altogether) the oral approach to history-keeping. In so doing, it changed our relationship to the past—and our relationship with time itself—which, in turn, changed the way we viewed ourselves as human beings. When history is told aloud, it has the effect of continually reinvoking and resacralizing past events; this keeps them alive in the present. The past keeps returning, for time is a wheel, ever revolving.

Stories and ceremonies have brought people together since time immemorial to celebrate the cycles of creation. In many cultures, they do so to this day. In this way, time is a living practice. But with the shift from oral to written history, the sensibility of living time has been largely lost.

The Western concept of time as linear, not circular, is making inroads into even the remotest cultures; everyone is beginning to feel colonized by time. This is true even within the West, where we did it to ourselves! In a certain sense we are the most colonized peoples of all, because we have internalized linear time so completely that we no longer recognize it for what it is—an illusion and a source of oppression. When time is conceptualized as linear, we organize events in our minds according to an imaginary timeline—much like the ones we used to study in elementary school called history. We put events behind us almost as soon as they happen because the past is past. "What's done is done." The past is no longer alive, no longer accessible. This way of thinking has consequences; it effectively severs our connection to natural cycles—which always maintain connection with the past, and with the wisdom of the ancestors. The latter were once believed to continue to communicate to the living through stories, dreams, visions, and even through the land herself.

But, you might ask, is the modern Western way of thinking not more sophisticated because it is based in a more accurate perception of reality? Not necessarily. This thinking is a product of a certain consciousness; it has been shaped by select ideas about the nature of reality. We call this *historical consciousness*—a way of thinking that emerged from the idea that events are separate, distinct, and irreversible. Historical consciousness has come to permeate our thinking to such a degree that *tense* (which has the purpose of distinguishing events in time) became an indispensable component of the grammar of modern Indo-European languages (including English). This was a pivotal development because tense and the perception of reality are now intertwined; we cannot communicate in English for long unless we know something of when a particular action took place on our imagined linear timeline of successive events. We have internalized this timeline so completely that it has become hidden from our awareness, and we no longer question the temporal aspect of consciousness now embedded in our language.

We have already discussed many of the ways in which historical consciousness tends to ignore the wisdom of the past, leading us to focus instead upon a future that is removed and abstract. But it is not only the past that is deadened; the present, too, is nearly killed. Why? In a linear view of time, the present is not an open doorway, but an isolated and fleeting point in time between past and future. In fact, if we are strictly linear about it, the present is an abstraction with no basis in reality at all because nothing we think of can be acted upon in the same moment. Our attempts to segregate time into sections are ultimately futile. Time does not fly like an arrow in a straight line; it flows where it is called to flow, in a manner akin to water. A linear notion of time is analogous to reaching one's hand into a stream to pull out one drop of water. It cannot be done.

To my mind, there is something suspect about a worldview that deadens the past, turns the future into an abstraction, and cannot grasp the present. Yet, the modern West is so sure of its view of time that we often hear the claim that it is an intuitive view of reality and therefore absolutely correct. The implication is that our conception of time is irrefutable and comprehensible to all, regardless of culture, language, education, or any other variable. Benjamin Whorf disagreed. He asserted that no absolute view of time or space exists; that these concepts are "recepts from culture and language."[14] They are a particular construction of reality—not reality itself. This makes sense to me.

Our conviction that we are right about the nature of time desensitizes us to the validity of other worldviews. We do not quite know what to make of Indigenous and ancient cultures, which originally had no linear view of time (but, rather, maintained calendars based on grand cycles). We dismiss them as mythical or primitive. They must be stuck in the past, we reason. They have not yet advanced to the correct view of time as a linear progression of discrete events.

It is as if we believe that ancient cultures had no idea of time passing altogether.

But ancient cultures, and intact Indigenous cultures, have never operated in such a manner; they have always been keen observers of changes in their environment brought about over time. In fact, it is rather obvious that modern Indigenous cultures are more aware of the ecosystemic changes in nature than modern Western cultures. This is because their lifeways have always depended upon a close connection with nature—so much so that the monitoring of natural changes became essential to their worldview and linguistic structure. In keeping with the ways of nature, Indigenous views of time tend not to focus on isolated species or things, but upon *the processes and interactions between them*. This is why Indigenous worldviews are well suited to harmonizing and synchronizing with the unfolding changes in nature—becoming one with the organic essence of time.

By contrast, our Western view of time is soulless and clinical, tied as it is to the Newtonian notion of time as an absolute, divided into equal increments. This construct has merits, of course; it enables the precise measuring of intervals of time. We have developed technologies that quantify time intervals down to nanoseconds, and we use this for specific human purposes, such as electronic trading on the stock market and other high-speed computer applications. (Whether these applications are genuinely beneficial to society is an interesting side discussion I refrain from entering.) The important point is that these intervals are abstract; they are man-made, not based upon the actual timing of nature.

For us, now, time is an abstraction, and our exclusive reliance on that conceptualization—while enabling some capacities—has also greatly limited our ability to see nature as it truly is: an organic, unfolding whole. We are particularly blinded to our place in that unfolding, for our linear view of time has pulled us out of the circular flow of the cosmos.

Our abstract view of time is reflected in our language; it is baked into our grammar, which is why English is replete with nouns. Nouns express a universe of objects; nouns imply a solid, stationary world. Again, this style of language has certain advantages. It enables us to isolate material objects from each other and to pin them down in a precise way, which is a contributing factor to why English has become the universal language of science. But this is also why speakers of English, and all languages of subject-object division, have difficulty expressing the flow of nature.

This crucial limitation, endemic to these languages, is a legacy of perspectival consciousness. It causes us to simplify reality, to see it, through our egoic consciousness, as made up of separate objects, while everything in between is just "space." It is a very static and frozen picture of reality. As a consequence, we no longer see or understand nature as an interconnected continuum, continuously moving and interacting, one thing upon another.

That is why, when we do succeed in immersing ourselves in living, flowing nature—for instance, when we raft down a river or gallop on a horse—we feel a profound sense of release (from our controlling ego). In these moments, we do not have time to think, and this feels liberating, even ecstatic. Yet, our reverie will be broken, sadly, as soon as we want to know *who* is having the experience. This very thought causes separation. It causes us to reimpose the "I"—which, in turn, divides us from what is being experienced.

It is in this way that languages have a surreptitious hold on us. If we were not thinking in particular languages, we might all experience nature very similarly, but because we are, our experience tends to be mediated through these linguistic lenses. It is not that a language causes a worldview per se, but language moderates the relationship between our perceptions of our lived experience as well as our expressions of, and reflections upon, that lived experience.

In this way, different languages have different gifts and drawbacks. For instance, Native languages are immensely rich in expressions of ever-changing processes and relationships. Their strength comes from their flexibility; they are supple and resilient, like a willow in the wind. Because they are primarily process-based, they are particularly suited for tracking the subtle shifts in the ecology of specific places. By way of example, Dan Moonhawk Alford famously related a conversation he had with Sakej Youngblood Henderson about the naming of trees, in which Sakej tells him that the Mi´kmaq name for a certain tree could be based on something as specific as:

> The sound the wind makes when it blows through the leaves during autumn about an hour after sunset, when the wind usually comes from a particular direction. So one tree is more like a shu-shu-something and another more like a tinka-tinka-something. [So] I asked him whether the name would change if something happened to the whole tribe of trees at once to change the sound—would the language keep up its fit with reality? Then that means, I said on hearing the affirmative, that if an elder remembers these trees over here were called such-and-such fifty years ago, but now after fifty years of acid rain they are called so-and-so, these trees could be seen as valid scientific markers for showing the effects of acid rain" [which Sakej confirms].[15]

Then, Moonhawk would invariably add: "And that's something we can't do with 'oak,' 'pine,' and 'mahogany.'" And he'd throw his head back and laugh out loud. His laugh cemented this story in my consciousness, for it helps to have a little levity if we wish to understand the serious limitations in our own worldview. Native languages are a precious resource because they are windows into a different worldview, "an old-growth forest of the mind."[16]

But Native languages also have their drawbacks, as illustrated in another story that Moonhawk liked to relate about Sakej. One time, Sakej was preparing to leave for an extended trip, and he gave elaborate instructions as to what was supposed to be done on his house while he was away, various repairs and improvements on the roof, front porch, and so on. But when he returned from his trip, nothing at all had been done. Sakej was disappointed and exclaimed, "Darn it, what we need around here are some more nouns to pin things down!"

English has no such problem. English, with its plenitude of nouns, is quite suited to pinning things down, prioritizing, and linear planning in general. This makes English highly appropriate for the exacting jargon of most scientific and business discourse. But when it comes to particular kinds of sciences, such as ecology, chaos, or systems theory—all of which attempt to articulate complex and subtle interrelationships in nature— English is ill-equipped. Noun-centered languages are limited when we enter the realms of incessant process and movement, as physicists discovered when trying to describe the quantum realm. To his credit, David Bohm realized this dilemma and went so far as to try to adapt English into a new mode of language, composed entirely of verbs, which he called the "rheomode," from the Greek *rheo*, meaning to flow. Bohm's experiment with the rheomode revealed some interesting implications for language as an undivided field of movement. It also exposed the tendency of English (as we ordinarily use it) to put "strong and subtle pressures to hold onto a fragmentary world view."[17] The implications of Bohm's experiment become even more interesting when comparing English with Native American languages, the latter of which are inherently more based in movement, process, and relationship. Native American languages see both time and space quite differently—but perhaps not so differently than a visionary physicist such as Bohm.

How Do Native Languages Express the Passing of Time?

With the exception of the rheomode as noted above, Indigenous languages express a sense of time that is completely different from that of Indo-European languages. The Native American Hopi, for instance, keenly observe changes in nature as expressions of a cyclic phase of becoming rather than discrete, unrelated events occurring on a linear timeline. Rather than seeing events as discrete and separate, they are seen as related and building upon one another, gathering momentum until they reach a full flowering at the end of a cycle.

In this conception of time, everything is connected; no past event is ever lost or without meaning. Further, repetition is never seen as tedious, as it might be for us; instead, repetition is seen as the way nature builds momentum. As Whorf explains, "To the Hopi, time is not a motion but a getting later of everything that has ever been done; unvarying repetition is not wasted but accumulated. It is storing up an invisible change that holds over into later events."[18] Think of a tree budding in spring. It happens because the tree has expended considerable energy repeatedly pushing up sap until it is time to bud. The Hopi recognize how nature works, how energy builds up until it culminates in action, like rainclouds before a storm.

For Indigenous peoples in general, the present is not a separate event, as we might see it, but a seamless part of the ongoing cycle of time as it manifests and unfolds. This kind of thinking lends itself perfectly to ecological stewardship, because this is how nature moves, in myriad recurring cycles. Repetition is nature's way of unfolding the whole, as we see in the fractal geometry of a seed head or pinecone. This is also why repetition is such an important part of Indigenous ceremony, since ceremony echoes Nature. By contrast, we (Indo-European speakers) are trained to believe that individual actions alone matter, and therefore we are impatient with repetition. This is partly why there is so much to be desired in our own ecological stewardship, because we find repetition boring and lose interest.

To put it simply, Indigenous people readily understand one of our proverbs: "Well begun is half done," but would have difficulty understanding another: "Tomorrow is another day."[19] The first implies that natural processes accumulate over time; the second implies that we can always start over, ignoring what has already occurred. This is one of the root causes of misunderstanding between Native and non-Native peoples today. For non-Native peoples, the atrocities of the past are over, but for Native peoples, they still reverberate, held in the vibration of the land herself.

The idea that the past is still happening is hard for most Western people to understand or accept. But why could it not be so? Imagine tossing a stone into a body of water and seeing it send ripples out in expanding concentric circles. These circles become fainter and fainter until eventually they are invisible to the human eye. *But they continue,* because all vibrations continue—they just become increasingly subtle over time. Everything in nature vibrates, and everything that has ever occurred is still vibrating and therefore still happening, at least in the subtle realm. The past is never really past; it is still occurring in the now.

Amazingly, such an understanding is contained in Indigenous languages. In Hopi, for instance, there are no equivalent words for past, present, and future. Instead, there are two realms we could call the "manifested" and the "manifesting." The manifest realm comprises everything that has ever existed in the physical universe. It makes no distinction between present and past; with respect to time they are both one. The second realm is the manifesting, that is, that which is coming into being but has not yet manifested. It includes what we refer to as the future, "but not merely this":

> (I)t includes equally and indistinguishably all that we call mental— everything that appears or exists in the mind, or as the Hopi would prefer to say, in the HEART, not only the heart of man, but the heart

of animals, plant and things, and behind and within all the forms and appearances of nature … in the very heart of the Cosmos itself.[20]

The astute reader will notice that there are strong parallels between the Hopi idea of time and Vedic cosmology as outlined previously. Both views recognize that what is yet to become manifest already exists in potential; both views recognize that manifestation exists first in the realm beyond our human senses. For both, time is a circle, not a line.

What Shifts if We See Nature as a Circle?

To Indigenous peoples, such as the Hopi, Navajo, or Blackfoot, all of nature, including the human being, is a great Circle. As we have just seen, the same belief is still prevalent among Asian peoples and, although we may not realize it, the same was also true for the ancient Greeks. This is an ancient, virtually universal understanding with much to teach us.

If we look at life this way, everything shifts. Life becomes an ongoing, continuously recycling creative process in which seed becomes root, bud, and fruit before returning to seed and starting the process again. It is a process in which life begets death, and death begets life, for death is only the fertilizer for rebirth. The seasons of one's life and the seasons of nature mirror each other. Ultimately, the circle of life is a process in which the invisible, or etheric, seeds of consciousness manifest in the physical realm, then return to the realm of formlessness, and then again becomes manifest, only to become reabsorbed into the formless flux of consciousness where they will later remanifest in their proper season. Thus, the circle of life repeats, but not exactly in the same way. It evolves—or unrolls—into a spiral, much like the structure of DNA.

The most interesting elucidation of this subject I have ever heard comes from Little Bear, who explains how, in the Blackfoot language, all the world is seen as massive fluidity emerging from and returning to an unbounded universal flux.[21] What is most interesting is that the Blackfoot do not see the world as being composed of bodies within bodies at all. As Little Bear and his Blackfoot colleague Heavy Head, explain:

Not all cultures generate a perception of reality as a fragmented landscape of solids within solids, acting as agents of change in the world. Some—like Blackfoot culture—produce experiences of fluid event manifestation, arising from and returning into a holistic state of constant flux.[22]

This conception of the universe as constant flux is very much in tune with modern quantum theory and has much in common with ancient Oriental notions of perpetual change (particularly in Taoism). But the Blackfoot worldview may take this idea the farthest.

To the Blackfoot, individual people are not the principal agents of change in the world. In fact, the whole concept of human agency is irrelevant to the Blackfoot. In simple terms, although the Blackfoot see the world through human eyes, the human experience is not bracketed and separated from the manifesting of all of creation. Instead, the experience of being human is completely embedded in the unfolding of all life events. This is only possible because, in a traditional Blackfoot view, there is no ego, no separate consciousness apart from the world acting upon the world. I find it rather remarkable that modern Blackfoot language continues to retain this way of seeing to any degree after five hundred years of contact with Western worldviews.

In Blackfoot, the human being is a part of the circle of life, otherwise known as the Sacred Hoop. Contrast this view with our own, in which humans stand outside—and above—the rest of creation. We consider ourselves the primary drivers of manifestation; we are separate *subjects* acting upon separate *objects*. But to be inside the circle in the Blackfoot way is to be *of the circle*, an integral part of the whole. In the circle, the subject-object division dissolves. This is why, in the dialogues, Indigenous people have sometimes spoken in terms of "I am the land," or "I am the water."[23] In Indigenous languages, there is no barrier (assumed or implied) between human and nature.

There is another subtle distinction in Indigenous language and culture. Things happen when they are *supposed to*—which, if you think about it, is how nature operates. Right now, as I write this, it is early spring.

Daffodils came up weeks ago, and now the tulips and soon the irises will bloom. The desert sage, chamisa, and the aspens outside our home have all greened up, but not the mountain aspens at their proper altitude, nor the willows and butterfly bushes. What seems dormant is really far from it, busy pushing up sap, preparing to bloom. Everything happens in its own timing. There are early bloomers and late bloomers, and all of this is contained within the grander cycles of life.

How Did the West Originally See the Human Being and Time?

It is important to realize that the Western view of the human being was originally very different from today, and so was our view of time. In fact, the two views cannot be separated. Time, or timing, was originally an integral aspect of being human. We lived in accord with natural rhythms, and we measured time in the same way. Time was real, in the sense that it was tied to the rhythms of the Earth, and knowledge of natural timing was essential to our survival.

But, gradually, we learned how to separate ourselves from time in order to manipulate it. This is how time became an abstraction, separated from origin. Our inventions tell the story. The first devices for tracking timing were sundials and astrolabes, which moved in synchronization with the rising and falling of the light. We then moved away from natural rhythms to measure time in an abstract manner, but we did so ever so slowly. Our first clocks and watches still approximated sunwise motion (what we call "clockwise" today, as if clocks themselves hold wisdom), but these devices represented a subtle step away from natural cycles. By Newton's day, we were already rationalizing this, reimagining the celestial spheres as a "Grand Clockwork." When, in recent times, we switched to digital watches (and digital everything), we seemingly abandoned all relationship with natural motion. But, because we can never completely let go of our relationship with nature, we now project the ways of our devices onto natural processes; we see nature as a computer, machine, or hologram. In so doing, we have reduced nature to a carrier of information,

rather than recognizing her as the primary creator and source of inspiration. This is not merely sad; it has dangerous consequences because we
are mistaking the imitation for the real thing. We are well on the way to
forgetting the true origin altogether.

We took a similar path with calendars. Originally, calendars were
based upon real lunar cycles. Some, such as the Mayan calendars, were
based on even more complex and exacting cycles of relationship between
the Earth, moon, and stars. But the Gregorian calendar is no longer in
tune with actual natural cycles. It has twelve *irregular* months rather than
thirteen *regular* months in accord with the lunar cycle, the latter of which
is also in keeping with women's menstrual cycles. A lunar way of keeping
time was not distant and abstract; it very much included the human being,
and particularly women, within the larger cycles of nature. Women were
once seen as natural wisdom keepers in part because wisdom was once
associated, as it should be, with the recurrent cycles of nature.

With the advent of the Gregorian calendar, we abandoned this original way of timekeeping. We embarked upon a different path. Once time
became separated from nature, it was a short leap to imagining that
human events could be separated from natural events. Once we separated from time, our separation from nature was inevitable because, ultimately, there is no difference. And separation became a force with its own
momentum, increasing exponentially. Over time, this separation from
nature became so complete that we began to identify ourselves by it. We
even began to separate ourselves from creation in our own thoughts; in
fact, this became a signature characteristic of human beings. This is why
we identified ourselves as the rational animal.

Yet, even today, it is possible for us to rethink our relationship with
time, nature, and thought itself. In fact, it may be imperative that we do so
in order to uncover what it truly means to be human. But to contemplate
this, we first need to still the mind—even if for just a moment—so that
true inspiration can enter.

> *"Is it the flag that moves, or is it the wind that moves?"*
> *"No," answers the Zen master,*
> *"It is the mind that moves."*[24]

Chapter 8

Beyond Rationality

What Would Be the Reality that We Would Experience If We Did Not Think in a Language?

This was the corollary question to the question of original thought asked by Little Bear in the 1999 SEED dialogue. Because it is even more challenging than the original question, I have refrained from (directly) addressing it until now. It becomes important, however, when we consider the impact that divergent languages have on human consciousness and human interaction. So, what would happen if we were not thinking in a language? Would there still be the same divisions between people and animals and between people and people that occur through language?

It is, of course, a koan, to think about "not thinking"; you cannot really do that—at least not successfully. But it is possible, if not easy, to quiet your mind. If you stop, or at least slow down your internal dialogue, a state of calm will follow. In Yoga philosophy, a peaceful and calm state of mind is known as *Kshina vrtti*. In this state, unnecessary mental activity is diminished, and the mind resumes its original form of *samapatti*. It is in this original state of mind that we recapture the creative essence of our being, giving us the capacity for fresh combinations of increasingly interconnected thought.

What we generally think of as thought is that which has already passed through the filter of language. But there is a moment—however fleeting—when thought/spirit is not yet language. In that prelinguistic moment, it is pure energy, unbound by rational mind. This is the phenomenon we call inspiration (from the Latin *inspirare* or "the flame of spirit being blown in") and it is most likely to occur when the mind is calm—when it is in its original state of openness. When inspiration comes, it is whole in itself; it tends to resist being broken down into component parts. It resists form, but is a noncorporeal essence or *archai* (original seed) of a form yet to come. It may prompt us to dance, paint, or sing long before it germinates into what we recognize as thought.

A sense of emptiness—no thought—begets fresh insight, perhaps even a form of the unmediated knowing that Krishnamurti referred to as "direct perception."[1] This may be as close to an original thought as is possible. Of course, the cultivation of an empty, receptive mind is almost inconceivably challenging when we are all the time so full of ideas. Little Bear likes to tell the story of the scientists who visit a Zen monk. The monk offers them tea and then proceeds to keep on pouring the tea until it is overflowing onto the floor. The scientists shout out in protest, but the monk simply turns to them and says: "Your minds are like these tea cups. They are so full they cannot possibly take in new information. Empty yourselves out and then come back."[2] In order to take in genuinely new knowledge, we must empty out what we previously believed—to see the world through the eyes of a child or, as they say in Zen, with beginner's mind. Beginner's mind is a mental state that dissolves illusory separation and embraces our natural state of interconnection. In reality, we are never separate from the source of our inspiration. It is only when we withdraw into our own thoughts, formulated in abstract, disconnected language, that we imagine we are separate from other people and nature.

Only in Our Dreams: Are Dreams a Form of Nonlinguistic Thinking?

The dreamworld seems to be a world without language, or without thought as we ordinarily describe it. And, as far as I can tell, all of nature dreams; our (human) dreams are only a part of a larger unfolding. The dreamworld is therefore an important starting point for exploring different forms of consciousness and the interconnection of humans with all of creation.

In my experience with dreams, while they are unfolding, I am not in the picture—at least not in the same sense as when I am in the waking state. The dream state is much more of a direct experience of undifferentiated consciousness. There is a consciousness, but it is not me; at least, it is not coming from or limited by my individual perspective. Rather, the dream consciousness is diffuse, spread throughout the imagery. But as soon as I awaken and remember the dream, I typically insert myself into the dream as the primary point of perception. In so doing, I impose upon the dream report a subject-object duality. Although this is virtually unavoidable in the English language, it takes away from the inherent wholeness of dream imagery. The dream comes in whole, and the imposition of a subject and object breaks it apart.

Eric Fromm has called dreams the "forgotten language."[3] But I am not sure they are a language so much as a way of seeing. Dreams seem to be a throwback to original consciousness, a consciousness before perspective. The dream ego is not the same as the waking ego; in my opinion we cannot speak about it as an ego at all. A dream is a form of direct perception that occurs without the interference of ego. And because dreams are not egoic, we should not be surprised when they do not follow the norms of our waking existence—including our conditioned beliefs about time. Dreams occur in their own unique form of time, which is remarkably fluid. After we awaken, we may say that the dream happened fast or slow, but it really happens at its own inimitable pace because dreams are not subject to the same constructs of time that we impose upon ourselves in our waking state.

Dreams are often "precognitive"—accurately predicting events yet to happen. There is much debate about whether precognitive dreams somehow *cause* the future.[4] But these debates are missing the point. The propensity for what we call precognition points toward something far more intriguing than an ability to predict or determine a single event on a linear timeline; it indicates that we are capable of accessing states of consciousness beyond that of our waking, rational mind. A precognitive dreamer is not seeing a future event that is somewhere out there, apart from them. Rather, she or he seems to be tuning into the subtle, inexorably unfolding forces of nature, of which the dreamer's consciousness is an integral part. And because dreams defy the limits set by our waking consciousness— such as time—they simultaneously open up the potential for creative thinking. Our dreaming self is highly creative; it braids together things that, in our waking state, we would dismiss as too disparate and unrelated to be taken seriously. But while we are dreaming, a dream makes sense. It is only after we awaken and try to fit our dreams within the confines of our rational mind that we find them to be nonsensical.

I believe that dreams—and visions—are examples of original thinking in that they are naturally inclusive, interconnected, and whole. They bring together in wholeness what otherwise would remain separate. It is not a coincidence that many powerful scientific and artistic revelations were initially experienced in a dream state—because the nature of original inspiration is completeness, and dreams arrive whole. Dreams bridge paradoxical and ambiguous dilemmas, which is another hallmark of creativity. Even dreams we dismiss as ordinary are often full of creative potential. Dreams connect us with our ancestors, with other cultures, with other animals, and with other dimensions that we are actually a part of, but (in our waking state) imagine we are not.

In a dream, our habitual, learned tendency to divide the world into separate, seemingly unrelated objects is suspended. The entire visual field is energetically connected and charged with numinous energy. This can create emotional turmoil that is difficult to handle, including what we commonly call "nightmares." We are told to ignore these strongest of dreams, but night *mares*, brought to us on the wings of a powerful horse, are perhaps the most important to remember. Why would the psyche try so hard to jolt us into awareness, going so far as to insist we wake up, if we were not supposed to remember the dream? Our dreams are a mighty river that must not be dammed up. If allowed to flow, our dreams can be a portal to domains far beyond those of our current philosophy.

My thoughts about dreams have been continually changing and evolving over the past three decades. Right after I moved to New Mexico, in the spring of 1994, I was blessed to meet and get to know a Navajo-Oneida botanist named Donna House who, as good fortune would have it, was my neighbor. Donna, who later participated in the SEED dialogues, was at the time consulting with the Smithsonian on the forthcoming Museum of the American Indian.[5] Donna and I used to have long talks, and the subject of dreams came up often because it was a shared interest.

At that time, I was actively involved in working with dreams as both a leader and participant in dream groups after learning a particular method from the late Montague Ullman, a well-known dreamworker who coauthored the book *Dream Telepathy* with Stanley Krippner. Monte once said that a dream was like a newly caught fish flapping around on the deck of your boat. If you are fortunate enough to catch one, you had better work with it all you can before, like a fish, it slides off your hook and escapes into its original domain. I had basically accepted this idea, but while talking with Donna, I suddenly realized that if I really wanted to understand dreams it would be better to jump into the ocean with the fish! For, as long as we take the dream out of its domain, pulling it up into our world for the purpose of analysis, we are basically dealing with a fish out of water. We can learn something by studying this representative of another domain, but if we really want to learn about it, we have to enter its domain.

One year at the dialogues, Linda Hogan told a true fish out of water story. Scientists had discovered a type of fish at the bottom of the ocean they had never seen before. To discover it, they had to shine a bright light on the depths of the abyss. We can only imagine how the fish, who lived in total darkness, must have felt to have a bright light shining on it. Then, they did the unspeakable and brought the fish completely out of the water so they might study it. Well, not surprisingly, it died immediately. A dream is similar in certain respects. It lives in its own special domain, one that balances the waking domain. If you take it out of its domain, and try to work with it in terms of the waking domain alone, it will be diminished; it might even die.

What Is the Relationship between the Dream and Waking Worlds?

Whenever I have looked at my dreams over a period of time, I have observed a certain consistency to them, as in a story that unfolds on its own terms. At first, this seemed to confirm the idea that dreaming was a domain all to its own. But, in recent years, I have started to notice that dream stories do not stop when I am awake; instead, they continue, like a background soundtrack, just below the focus of my everyday attention. I am beginning to see the dreamworld and the waking world not as separate domains that occasionally overlap, but as complementary domains that interpenetrate each other at all times.

I have come to realize that one of the most powerful things we can ever do is to stop dividing our day into sleeping and waking. For, in truth, they are two faces of the same experience. In writing this book, whenever possible, I work on it during the day until evening, and then before going to sleep I read what I have written. Then I continue where I left off first thing in the morning. I rewrite what I wrote the previous day upon awakening, because I know (even if I do not fully remember my dreams) that my nocturnal experience will add depth to my waking vision.

Our dreaming self does not stop when we awaken; if we give it breathing room, it can bleed through into our waking state and offer guidance. Conversely, our waking consciousness does not disappear when we sleep; it feeds our dreamworld in much the same way. If we dissolve the dichotomy between our waking and sleeping selves, these seemingly separate awarenesses become two rivers that flow into and out of each other. As the Mayan-trained healer Martin Prechtel said, our dreaming and waking selves are two wings of a butterfly. Neither wing contains all of life. The butterfly must have both wings connected at the center, its heart, in order to fly.[6]

What Is Real?

Chuang Tzu famously asked, "Am I a man dreaming I am a butterfly, or a butterfly dreaming I am a man?"[7] Which, or what, is real?

We may find it hard to unite our dreaming and waking selves. Many of us dismiss the dreamworld while taking the material world as the full reality; others feel less at home in sensual reality, ignoring it to daydream and lose themselves in imaginary musings. Both are extremes, and I believe we should aim for integration.

Much of our waking-day reality is conducted in an unconscious haze, and we too easily take whatever we are thinking as literal truth when, in fact, we are seeing only a small part of the entire picture. Paying attention to our nocturnal dreams can, I believe, profoundly expand our scope of awareness. So can cultivating our capacity to bear witness to our own thought processes. We need to pay attention to what we are thinking and observing while we are awake *as it is happening* and, if possible, try to recognize what it is we are habitually dismissing before it passes out of our conscious awareness. This book discusses thinking processes in greater detail in Part Three (Return), but the point now is that paying attention to our waking thoughts and our nocturnal dreams—these two practices—can both help reconnect us to the larger consciousness that is all around us.

Becoming Truly Human: Are We Fully Human at Birth or Do We Learn How to Become Human?

In one of the early dialogues, Little Bear began with an intentionally provocative statement. He said that fish are more evolved than humans because fish know how to be fish. The implication was clearly that humans are still learning how to become human, which may explain our fascination with our own humanity.

Little Bear's statement gives me pause. It is interesting that, of all creatures, he chose to compare humans with fish, because human life, as all other life, is thought to have originated from the sea. Indeed, we still begin our lives in the watery environment of the womb, and an embryo appears more tadpole than human, only later developing into primate and then human form. As adults, we are 70 percent water, but at birth it is even a higher percentage, about 90 percent. Water even provides our principal means of seeing the world, for the human eye is filled with liquid. And when our watery eyes first behold the world, what do we see? I believe we see oneness—that we are born undivided from spirit.

When we embark upon our life, our consciousness is diffuse and inter-connected, much like a dream state. But those of us raised in modern Western society quickly learn to change that consciousness as we are taught to individuate, to differentiate ourselves from others. We continue to distinguish ourselves from others, along with differentiating all kinds of objects from each other, well into our adult lives. Most Western psychologists think this is both advisable and universal—that all humans must develop their individual egos *before* we can begin to learn the interconnections that bind us together as human beings.

But ancient peoples had a different philosophy and, thus, a very different way to begin life. The newborn was understood to emerge from a greater spirit world, and there are many birthing traditions based on this assumption. These traditions are fully or partially intact among many Indigenous peoples around the world.

Among the Dagara people of Africa, for example, a birth is known as "welcoming spirit home."[8] The entire process begins well before pregnancy with a fertility ritual. This ritual takes place in the time known as "spiritual pregnancy."[9] A welcoming state of mind has already been cultivated in order to attract spirit into the world. When a physical pregnancy occurs, there is a celebration; and, at the time of actual birth, there is another celebration with the whole community present to welcome the newborn into the world. Significantly, this community is not just the biological family, and not just the village—but the whole family of human and more-than-human relations, including the human ancestors and the grandfather trees, the plants, rivers, mountains, rocks, and the very soil of Earth, the mother of all. Everything and everyone is called upon to welcome the approaching birth—to welcome spirit home.

Children are a particularly important part of the welcoming committee because children are an honored part of the community. Their presence at birth helps impart a sense of interconnected community to newborn babies. The intergenerational presence of the human community is coupled with the sounds, smells, and fresh air of nature—all of these factors come together to facilitate an emergence into the world that is far more seamless than we moderns typically imagine. For instance, it is not uncommon for the newborn infant to crawl up onto their mother's breast—something we moderns, obsessed with clinical birthing, find hard to believe.[10]

Sonbofu Somé, a native of Dagara, has taught many of these traditional birthing rituals to the modern West and has also developed birthing ritual techniques that can be utilized even in a hospital. When a birth takes place in a hospital, it is far less likely that the baby will spontaneously crawl up to their mother's breast, but Somé advises mothers to ask for the infant to be placed on their breast before disconnecting the umbilical cord. She also suggests that recorded sounds of children, nature, and relatives be brought to the birth site.[11] These practices are beneficial for newborns and the adult community alike because, to the extent that these ancient rituals are followed, we start life on the right path—the path of reclaiming both our interconnectedness and our sense of what it means to be human.

How Can We Recover Our Original Sense of Community?

I believe that human beings have a purpose, and that our purpose is to learn to be fully human in harmony with the rest of creation. But we cannot learn this on our own. We learn through relationships—by making alliances with the many other beings, both human and not—that participate with us in the journey of life. This is something we used to know, but have unlearned. It is something we can know again.

Our community used to include all of nature. We were once connected, not just to our fellow human beings, but to—and through—the air, water, and soil of our place of origin. We were connected through the central fire around which we gathered to tell our stories. That fire, in turn, was connected to the fiery sun, and all the stars that enfold us. We were connected through the moon and the tides and the ever-renewing cycles of night, day, and the seasons. Nothing was considered lifeless or excluded from the sacred circle of life; all was essential. This included the swirling winds that blew through the lands and through our lungs and the waters that snaked through the canyons and through our veins. And it included the grandfather rocks that literally held a great position in the community.

In the same way, the rooted trees, upon which grow the mosses and lichens that feed the insects that feed the birds and then the mammals—many of which we once hunted—are all a part of our community. Around the tree roots live the plants, bushes, grasses, fungi, worms, and other microorganisms that aerate the soil in which we plant our crops. The trees and other vegetation provide us with a continuing supply of fresh air and a link to water, mineral, and vitamin sources—both food and medicine. The mountains similarly point us to the interconnected sources of water—the rivers, lakes, and oceans that we and all the other creatures could not live without. The mountains gather the clouds that bring the rains that feed the soil that runs into the underground waters and aquifers and out to the rivers, lakes, and oceans, only to evaporate and gather once again in the skies.

Living in the arid Southwest as I do, I am often reminded of the water that is abundantly present in the clouds, first congregating in the skies over the mountains and then gradually spreading out over the valleys. It is easy to see why mountains are considered sacred as they stand like guardians in each of the four sacred directions surrounding and watching over the valley communities. There is a great diversity of life in mountain ecosystems.

Once we learned from all our plant and animal relatives, for we knew them to be both kin and teachers. In the high desert of the Southwest, the bear is an important animal who demonstrates humanlike qualities, such as jealousy, fear, anger, wisdom, and introspection, and is also associated with the cyclic change of the seasons.[12] We learned much from the bear, but we have also learned from the trees, plants, and even the rocks, as well as a myriad of other crawling, flying, finned, or four-legged creatures.

The two animals that inspire me the most are the bear and the eagle. Accordingly, I keep these two "Zuni fetishes" (small carvings from Zuni pueblo) on my writing table. The bear and the eagle work together, for the bear is grounding and wise and a good companion to the soaring, visionary eagle. But the eagle is the most powerful totem for me. Eagles appear, in the wild, or in my dreams, seemingly whenever I need a larger perspective.

I once prayed all night in quest of a vision, and happily went without sleep. In the morning, as I gathered with others to welcome the sun, an eagle flew toward us from miles away. I remember seeing the eagle in my mind's eye even before it appeared in my field of vision. It flew with purpose, across the vast mesa, the nearby canyon, and directly toward us, in the direction of the rising sun. Then it passed overhead—still high in the sky, but clearly delivering a message. Because I had stayed up all night praying, I was in (what we in the West call) an "altered state," and I questioned for a moment what I had seen. At the first opportunity, I asked Alvira, the eldest daughter of Grandfather Leon, if that was an eagle I had seen. She was impressed that I had seen the eagle because apparently nobody else had, and it was most definitely an eagle, she assured me, for our mutual friend Susan Stanton had been gifted with a plume dropped in her lap.

Whenever I see or dream of eagles, I know a vision is coming. Eagles open me up to new possibilities, teaching me to trust in the larger picture of which I only see a part. It is amazing how powerful the eagle is in my own consciousness—in all of human consciousness—even as so few of us encounter them in close proximity. I relate to the eagle as a wise elder or ancestor—and sometimes, I believe, *it is the elder and/or the ancestor.* In the story I just related, I was praying on the very spot where Grandfather Leon had prayed, told stories, slept, and dreamed. It is a power spot soaked with his energy and that of many other mystics and seers before him. It also happened to be exactly three years since he had passed to the spirit world. I was praying to Leon, and I believe it was Leon himself who delivered the message.

At minimum, the eagle is a messenger who delivers insight from realms beyond where humans can go (unassisted by technology). The vision and oversight of eagles and hawks makes them a great ally and natural extension of the human community.

When we define community as a strictly human enterprise, we lose something precious and demean all of life. I am convinced this is why a lack of well-being pervades modern life. I used to wonder why and searched for the cause, but I came to believe our malaise is not because of any one factor: not our dissatisfaction in our interpersonal relationships; not the demanding nature of our work life; not any of the standard stresses of modernity—not the traffic, bureaucracy, or even the financial pressures. These factors are real, but they are symptoms, not cause.

The direct cause of our suffering is the rejection of our relationships with the natural world. Nature has become the other—something we have cast out from our life, or try to—because it is impossible to remove ourselves from nature because we are made of the elements of nature. All we have done is to make certain parts of ourselves unwelcome—our so-called animal nature—our gut feelings, our erotic yearnings, our desire to connect with and be one with a larger force that pulses through all of existence. We pretend to have tamed ourselves—to have become good girls and boys. In the process, we redefine nature as something untamed—something wild.

Human culture, in contrast, has become a euphemism for something refined, sophisticated, and distinctly unwild—like the symphony or ballet. But this is misleading. The word "culture" originated with "agriculture." Culture is a living organism. It grows, like live cultures in yogurt. It begins at the grassroots level, and organically develops. It needs to be cultivated, not tamed or repressed.

The refinement of human culture, however sophisticated, can never completely supplant its wild beginnings. This is why the wild side of me entertains a fantasy. Would it not be great if, someday, in the middle of a classical symphony, a bluegrass fiddler like John Hartford were to suddenly leap out of the orchestra and break into a hot version of "Orange Blossom Special" or "Lonesome Fiddle Blues?" I doubt it will never happen, but it is still something I would love to see.

Human community and human culture emerge from the earth and are vital, living forces. Our so-called human nature is not fixed; it is nourished by the soil of culture until it germinates and flowers. Thoughts, hopes, and aspirations act in the same way. They germinate like germs germinate. I use the word "germinate" intentionally because we are more like germs than we acknowledge. In fact, our existence is dependent upon germs, mostly helpful bacteria that live inside us, roughly 3 percent of our body weight.

To be human is to be inextricably connected to the community of life—and that means the whole community. As Aldo Leopold wisely counseled:

> Conservation is a state of harmony between [humanity] and land. By land is meant all of the things on, over, or in the earth. Harmony with land is like harmony with a friend; you cannot cherish his right hand and chop off his left. That is to say, you cannot love game and hate predators; you cannot conserve the waters and waste the ranges; you cannot build the forest and mine the farm. The land is one organism.[13]

Leopold fought for the reintroduction of wolves into the ecosystem, and thanks to him, today, there are wolves in the Gila wilderness of New Mexico. But I wonder today if he might fight today for a different cause: for the bacteria in our depleted soil. I sometimes think our war on germs is equivalent to our so-called war on terror. Both are futile attempts to root out something that actually lives within each of us. Rather than projecting our shadow of insufficiency out onto the world, we need to embrace all parts of ourselves. Balance and harmony requires wholeness. No part of the ecosystem is less than any other part. All our relations means all our relations. To be human is to be in relationship with every other creature.

Chapter 9

A Tale of Two Directions

I have sought to illuminate two contrasting views of what it means to be human. We may broadly call them Western and non-Western, although in truth the lines are no longer defined by geography.

The Western view believes that human beings are defined by their differences from the rest of creation. The non-Western view defines humanity mostly in terms of connection with the rest of creation. In effect, these two belief systems represent two different stories about the path of humanity. One is the story of modern science—of humanity forging its own direction based on human ingenuity and the march of progress. Nature, in this story, is to be overcome, outwitted, or controlled. The other is a timeless origin story, in which our human nature is derived from, and continually renewed by, interactions with nature, for nature is part of us—our relation.

These two stories present us with very different views of life and, thus, very different challenges. One of the most respected and visionary scientists of the twentieth century, Rachel Carson, understood the choices these different stories represented. She said:

> We now stand where two roads diverge. But unlike the roads in Robert Frost's familiar poem, they are not equally fair. The road we have long been traveling is deceptively easy, a smooth superhighway on which we progress with great speed, but at its end lies disaster. The other fork of the road—the one "less traveled by"—offers our last, our only chance to reach a destination that assures the preservation of our earth.[1]

Remarkably, the ancient Hopi rock carvings (or petroglyphs) at a site known as Prophecy Rock depict just these two stories. I was blessed to visit Prophecy Rock in a group escorted by Hopi elder and pipe carrier Jerry Honawa, who explained the meaning of the prophecy in simple terms.

As you can see in Figure 6, there are two distinct lines carved into the rock. According to Honawa, the long line charts the path of the original human being, Maasaw (A), who lives in tune with nature. His only tools are a gourd of water, white corn seeds, and a stick he uses to plant the seeds with. The peoples that follow this path are said to continue indefinitely (along the line marked B–G).[2] (The line in the image appears to end at G but, in actuality, continues around a bend in the rock not visible in this photograph.) The shorter line (C) charts the path of another faction of humankind, one that falls in love with technology and the manipulation of nature. This path ends abruptly in a time believed to coincide with our present day (F).

The Hopi believe that humans have already lived through three previous worlds, each of which was destroyed, and that we are now living in

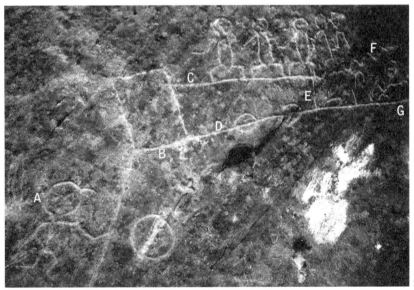

Photo: Jerry Honawa

Figure 6: Prophecy Rock

the Fourth World. It is in this Fourth World that we are being presented with the choice of which path to take. Note that on Prophecy Rock there is a connecting line leading back from the destructive path to the original path (E). This connection may represent an opportunity for humankind to return to the original path, to realign with our original purpose as human beings which, as we have been asserting, is to continuously learn how to live harmoniously with all of nature. It is unclear, however, whether this window of opportunity has already passed, or is still available. Being an optimist, I believe there is still time, as long as we change our thinking. I truly believe that all things will again become possible once we break out of the illusion that we are separate from creation.

There is no doubt that this is a critical time in the history of humanity. And, increasingly, humankind is realizing we must reevaluate our path, that our own reckless behavior has put our very existence—and the existence of countless other species—at great risk. More and more people realize that we must change our thinking now, or risk extinction. This is no longer an inflammatory or controversial assertion.

To change our thinking, however, we must change our story, or rather, reawaken to the original story of how human beings were intended to live. If we are able to do this, we may find that the story of modernity, as divergent as it has been, is coming around full circle to reunite with the stories of original peoples.

I say this not out of fanciful or wishful thinking, but from experience. I have been blessed to participate in dialogues that brought Indigenous peoples together with leading edge Western scientists, primarily quantum physicists, who are rediscovering a common origin—a cosmos of radical process and relationship. The funny thing is that modern physicists had no intention whatsoever of reuniting with Indigenous peoples when they pursued the subatomic realm. Why would they? Almost all modern physicists assume the universe is mechanical at its core, which is why the professional field is called quantum mechanics.

But, for a small minority of physicists, the study of the subatomic realm changed everything. Why? It could be because the quantum realm is the source of energy—the immutable source of vibration that governs everything there is. And energy, as physics tells us, cannot be created or destroyed. This is why we might also call what is inside the atom spirit or original thought, for quantum energy is inclusive, interconnected, and whole energy. It is life force—not intended to be broken apart—and we have seen what happens when it is: the atom bomb.

To discover our interconnection with all there is, we need not create the most destructive weapons ever imagined. We only need recover our original way of thinking. In doing so, we will regain our connection to all of life and our compassion for everything that exists.

> A human being is part of the whole, called by us Universe.... We now experience ourselves, our thoughts and feelings as something separated from the Rest ... a kind of optical delusion of our Consciousness. This delusion is a prison for us. Our task must be to free ourselves from this prison by widening our circle of compassion to embrace all living creatures and the whole of Nature in her beauty.
> —Albert Einstein

All Is Alive: From Separation to Right Relationship

In a certain sense, it seems obvious that, as a species, humans are the most separate from the rest of nature. But in a larger sense, it is untrue; it is only that man thinks he is separate from nature. No creature is separate from nature, regardless of what we imagine. All creatures derive their consciousness from the same source.

Everything that exists vibrates; vibration is life in its original and most inclusive sense. Thought is a particular expression of vibration, but humanity is not alone in thinking. All of nature thinks. Humanity is intelligent, but only because all of nature is intelligent; humanity is alive and aware only because nature is alive and aware. It may seem that thoughts arise within us on their own, but they are actually coarising from the ground of being, just as Buddha taught. Human consciousness is a reflection of the consciousness of all. What would it be like if we embraced this?

I began this book with a prayer because I have learned from my Native American friends to be thankful for the gift of life, and specifically from Grandfather Leon to be thankful to the elements because we are made of the elements. These same elements make up the entire cosmos, and unite us with all of creation. The elements are the crux of what it means to be a human being. We are alive because of the elements: *water*, *air*, *light*, and *earth*. The elements *are life*, not just the constituents of life.

Consider air. The breath (for the ancient Greeks *pneuma* and the Jews *ruach*) has been thought for centuries to be the carrier of spirit. If we think of breath as shared spirit, the air itself becomes aware—the medium of life—or, as the ancients called it, the *aether*. The aether was not a void of empty space, but a plenum of fullness; it was not a medium of separation, but a medium of interconnection. The Navajo have a similar concept of air (*nílch´i*), which I learned from Maryboy and Begay. The *nílch´i* is the same "one air," whether it exists outside the body or as a breath within the body. The *nílch´i* defines us as human beings because we are what we breathe in, including not just oxygen, but the totality of all existent atmospheres and gases that accompany each breath. The *nílch´i* is all encompassing, a part of the totality of cosmic process. It is within all processes on Earth and within all processes in space, including the black, blue, yellow, and white stars that provide a link to our ancestral origins. Our consciousness and cosmic consciousness are one, linked through the same *nílch´i*.[3]

This type of thinking underscores the reality of how humans are related to all things, irrespective of linear time or space. It is this expansive form of thinking that I call original, because it challenges me to think of every aspect of creation as radically interconnected, inclusive, and whole. Such thinking is, however, at odds with today's mainstream thought, which is still married to rational separation. In mainstream thinking, the elements are the constituents of life, but not alive themselves. Does this really make sense—is it really scientific? What makes the air we breathe to keep us alive all of a sudden dead as soon as we exhale it? And how could the air we breathe be dead, when it keeps us alive?

Everything that exists is alive and interdependent with us. Everything that exists combines together to continually create a dynamic, living whole.

Human interdependence with creation is most obvious when we consider our relationship with plant life. It is a biological fact that humanity is completely dependent upon the plant kingdom for our very breath, just as plants are dependent on the animal kingdom for theirs. We breathe out the carbon dioxide that they breathe in; they convert the carbon dioxide into oxygen that we breathe in. Breath is a circle that unites us.

With such complete and total interdependence, why should we be surprised when people report communicating with plants? A psychic plant-human relationship is but a "garden-variety level of telepathy," quipped Moonhawk, who liked to play with words. Human-plant telepathy used to be commonplace and is only natural, because we are bound together in a literal conspiracy in that we conspire, or breathe together. In this view, plants and people think together, and each can offer medicine to the other. Once, we openly celebrated a familial relationship with the plant kingdom, and Indigenous societies all developed significant medicines from plants. Traditional plant medicine works because human and plant are interdependent, kindred spirits.

This is one of the reasons why ceremony often includes the use of plants or plant medicine; this connects us with our plant relatives and, through them, to all of life. One of the most important ceremonies, at least for Native Americans, has traditionally been the pipe ceremony. I have also found the pipe ceremony to be deeply meaningful, and the best means I know for relearning what it is to be human. I therefore trust the reader will find this description a fitting way to conclude our discussion.

The Pipe Ceremony

The paradox of being human is that we are simultaneously of the earth and standing out from the earth, reaching toward the heavens. Because of our complex nature, we often get out of balance, favoring one extreme or the other—our earthly, material ways or our airy, more spiritual inclinations—forgetting to integrate these disparate parts of ourselves, and forgetting to restore a right relationship with all of creation.

The practice that best reconciles this paradox for me is the Native American pipe ceremony. I have had the privilege of participating in pipe ceremonies, and although there is still much I do not understand, there is much the pipe and pipe carriers have taught me. In particular, I learned from the late Tobasonakwut Kinew, pipe carrier, former chief of the Northern Ojibway (Anishaanabe) people, and a member of the Midewin healing society.

As Tobasonakwut explained, the pipe is representative of the human being. Traditionally, the pipe bowl and pipe stem are carried in the same container but wrapped separately. The stem (masculine aspect) is wrapped in red flannel, the color of protection. The bowl (feminine aspect) is wrapped in deerskin or some kind of fur associated with Mother Earth. When the stem and bowl are put together, this represents the joining of the feminine and masculine aspects of a human being. As human beings, we stand on two legs out from the earth but are always connected to Mother Earth, just as the feminine and masculine aspects are never truly separate but always found together in nature.

In preparation for the ceremony, the stem and bowl are both smudged for purification. (For the Anishaanabe, the traditional smudge smoke comes from burning sweetgrass and cedar, not the sage common to tribes in the desert Southwest). Tobacco is itself sacred for a great many reasons. The old stories say the Creator gave it to the people for the purposes of ceremony, and it is found far and wide on this continent. Like all plants, it has its roots in the soil of the earth and requires light, air, and water, elements both from the sky and the earth, to grow. The fact that tobacco can be smoked brings in the transformative element of fire, related to the light and warmth from the Sun, the source of all life on Earth. Tobacco's roots go deep into the earth from its *rhizome* or horizontal plant stem. (Tobasonakwut used the exacting botanical term.) These millions of roots are associated with a deep and long memory of the ancestors. That is why tobacco is referred to respectfully as Grandfather Tobacco.

The smoking of tobacco is related to other fire ceremonies. In ancient Anisshannabe traditions, the Waabano (or People of the Dawn) had many healing practices connected with fire. Some of the earliest ceremonies utilized water and plant material on heated stones—like a miniature sweat lodge—to heal. A person's head was covered with bearskin, and the ill person would inhale the medicine so that the vapors would go directly into the bloodstream. In a similar way, when tobacco is smoked, the smoke directly enters the bloodstream, and is thought to protect the veins and internal organs, the heart and so forth. The fact that modern medicine thinks the opposite today should not be too easily accepted, because many things we consider poisonous are actually medicinal in the right dosage.

In ceremony, when the tobacco is smoked, the smoke itself becomes an ally, called upon for purification, protection, and communication purposes. If you watch smoke carefully you see it does not travel a straight line, but a curved path. Eventually, it goes upward to realms beyond the reach of earth-bound humans. In this way, it is said to connect with the ancestors or other allies such as eagles, hawks, and other spirit beings who reside in these upward realms.

The ceremony includes the saying of prayers while the tobacco is being placed in the pipe. All six directions are acknowledged: the four cardinal directions and the above and below. By saying prayers to the Creator in all the directions, you are blessing the tobacco. After doing all these things correctly, Tobasonakwut said, those in the pipe circle can then speak to Creator directly. The pipe is then passed from person to person. In receiving the pipe, you hold the pipe with your left hand, and place it on your heart. This is another way to heal your veins and internal organs. Repeat the same for your forehead, so only good thoughts will guide you.

We inhale the smoke into our body, and when we exhale, we do so to all the directions. All of this should be done in a prayerful manner. The pipe ceremony is not a ritual, but a living prayer, a vehicle for transporting our thoughts in a sacred manner. Our prayer becomes powerful through the strength of our intention. If our intention is pure, the smoke will carry it to our entire lineage of ancestors, both human and more-than-human. In this way, our prayers are a form of original thinking, for they can connect us to allies and powers beyond the human realm.

The pipe ceremony continually reminds me of my connection with all of creation. It teaches us about our essential nature, which is paradoxical: the human condition is that we are both connected to source, yet ever diverging away from it. Thus, we need practices that acknowledge this and help bring us back into balance. It is in the act of seeking balance that we rediscover the truth of our nature, even as we never achieve a final balance.

The pipe is a powerful teacher of what it means to be human, and it is a teaching that is never finished. Like any ceremony or life practice, it is ongoing; our learning must be constantly renewed. To keep relearning who we are, to continuously seek to rebalance ourselves—this, to me, is an essential part of what it means to be human.

PART THREE

Return: How Has Our Thinking Created the World Today, and What Is Now Emerging?

The major problems in the world are the result of the difference between how nature works and the way people think.

—Gregory Bateson

We are not blind! We are men and women with eyes and brains … and we don't have to be driven hither and thither by the blind workings of The Market, or of History, or of Progress, or any other abstraction.

—E. F. Schumacher

Corn seeds, time, human culture, food, and all existence were all the same thing: Time was a plant grown from a seed; human culture was a seed; their food was a seed; all existence came from seeds; they were all seeds, and they were revered collectively as the living Holy … at all costs "keep the seeds alive."

—Martin Prechtel

Anything worth thinking about is worth singing about.

—Bob Dylan

The King

You may have heard about this king, some years back, who commissioned a map of his dominion, of his lands. He loved this terrain, but really wanted a fine map and he did a contest and chose the finest cartographer in the land, and he said, "I want the best possible map of this whole realm." And the cartographer took this very seriously, had various workmen and workwomen working with him and they surveyed, measured, and crafted with such art and so much beauty and with a lot of the finest pigments, a map—it had quite a large size it was like you know, six feet by six feet, that they brought into the palace, and unveiled it before the king who was delighted except ... "Wait, where's well, where's those mountains over there?"

"Oh, your Highness, those aren't ... you mean those hills?"

"No, no, that mountain right over there."

"But, your Highness, they're not big enough ... Look we've got the Alps over here, and over there are the Pyrenees and, but those are not big enough to really get on the map."

But the King said, "That is my favorite mountain range! I've been gazing at them everyday, I need them in the map."

OK. So the cartographer went back to his workshop, gathered his helpers, and set to work. And months later brought back to the palace a map, but this map was of a very different scale. In fact it filled up much of the room wherein the throne was, but they propped it up as best as they could. It was about 25 feet by 25 feet, and they said, "Look, your Highness, and here are those golden mountains, we named them for you."

And the King said, "Ah, fantastic! But where's that ... river. [gesturing out the window] come, you guys see that river over there? I don't see it on the map. I need."

And they said, "Your Highness, that is not a river. That is a creek. And it does not have a place on our map because it's not a real river."

"What do you mean it's not a river," said the King. "That river blesses me each day when I go down to it, and I take its water and drop it upon my neck, and my eyes, please ... that river."

But they knew, and they argued hard. They said, "Your Highness, it's not even a stream. And it runs dry every year, it's a little creek!"

"It's a river," the King said, "I've always thought it's a river, it's my river, it's the river. I need it on the map."

OK. So the cartographers went back to the workshop, and they worked and they labored, and they labored, and they came back months later and said, "We've done it, King."

And he said, "Bring it in."

They said, "We can't bring it in, you've got to come out here." It was the size of the palace itself. And he was able to step onto this map, and it had his river and it had his mountains. The King was pleased, but then—it didn't have the house where he was born—and they went through the whole thing again! They said, "Well, that house is … you were of humble beginnings, your Highness!"

"But that house is central to the world! It needs to be …"

The long and short of it is that they went back to the workshop, but this time it took them a couple of years because this is a pretty small house. When they were done, it wasn't just a couple of years, it was two and a half years or so, maybe longer, because they finally, finally constructed a map that satisfied this King precisely. It had every nuance within the land in it. It was the size, unfortunately, of the land itself. And so to lay out an unraveled map, they had to clear the entire realm and lay out this map over the landscape. But it had every spot, every stone, every stream, every dry streambed, every tree, and every dead snag was there. It was "the map." And the king and his followers were overjoyed, and they settled back in on top of the map, and lived there as did everybody else in the kingdom—until the rainy season came, and … the clouds gathered overhead and it started pouring … and the colors began running a bit on the map, which got a little frayed in some areas where it got overly wet … and it pulled where there were little slumps in the map.

And those places kind of just tore a little bit, and right through those various tears and thin spots in the map, something was peeking up—little bits of green, grasses, small plants, and seedlings, and people were looking at those and saying, "What's that?" "What's that stuff?" And more and more were appearing as pieces of the map detached themselves from other pieces, and these little bits of earth and soil, dirty soil, was appearing and smudging the edges of those pieces. And people were saying, "Hmmm, what's this stuff?"

Seems to me that maybe that's where we are today. The map is fraying, and we're beginning to—I hope and pray—discover something that's been hidden by the map itself.

—David Abram, told at 2004 SEED Dialogues

Chapter 10

The Essence of Thought

How Has Our Thinking Created Our World Today, and What Is Now Emerging?

The modern world is faced with countless dilemmas of our own making, all traceable to our thinking processes. It is not easy to change our thinking, however. We have already seen how ingrained are our beliefs about time, history, progress, consciousness, rational thought, and what it means to be human. This part of the book broadens our inquiry, examining how our thinking has created our contemporary institutions, including professions like economics that wield such great influence. We also deepen our inquiry, probing increasingly closer to the origins of thought and tracing the whole process of thinking from origin to manifestation in action. This is necessary to do because thought matters; thought is real in itself and has real consequences that affect people and nature. As Krishnamurti noted, it is our ignorance of our own thinking processes that create much of the sorrow, misery, and confusion in the world. Conversely, I believe that awareness of our thinking processes can result in deep structural changes in perception that can then give rise to whole, wise, and effective action in the world.

For thinking to be most effective, it has to be original in the sense of connecting us to a larger whole. It must be more than mere connective tissue, however. It must be generative, working in the way nature does, always optimizing, never compromising the whole. Our task in this book and, more importantly, in life, is to recognize this kind of thinking—thinking that comes from Source and enhances life—and then water these seed thoughts within ourselves and within others so that they grow and produce fruit.

At the same time, we must learn to get out of the way and allow these thoughts to mature on their own because ultimately these thoughts are not ours to possess. They come from a greater source—an impetus beyond our understanding. Our thoughts are like our children. They may be with us, but they are not ours (or ours alone). Gibran wisely counseled, "Your children are not your children. They are the sons and daughters of Life's longing for itself."[1] If we try to control our children, bending them to our personal will, we fail. The same thing happens when we try to consciously manage thought with our intellect. The result is hardened habits of thinking that become impervious to change.

Most of us are already deeply conditioned to think along certain lines. We might want to change our thinking, but it is extraordinarily difficult to do so. This is mainly because we unwittingly try to change our thinking with the same kind of thinking we now use. Even an apparently open-minded person who is willing to entertain a challenge to their assumptions will tend to interpret that challenge through their same habits of perception.

When Einstein said, "We can't solve problems with the same kind of thinking that created them"—a quote often cited because it is true—he did not specify how to go about making change or how hard it would be to think "out of the box." It is difficult to go outside one's paradigm because the nature of paradigms is that they inhibit new perceptions. This is a conundrum; it is hard to overcome, but perhaps not impossible.

To that end, I next offer a radical revisioning of thinking and paradigms that may, to some, seem unusual or even alien at first. But, if we are to break out of the box that holds our thinking hostage, this may be exactly what we need.

A Radical ReVisioning of Thinking: What Is a Paradigm?

We have already spoken about paradigms, but we have yet to give a working definition of the term. Ironically, this is precisely how paradigms operate; they are assumed to be understood, so are not made explicit.

Paradigms, and Western epistemology in general, tend to isolate ways of knowing by definition—with profound consequences. The word "paradigm" comes from the Greek *paradeigma,* meaning "pattern," the Greek *paradeuknunai,* indicating "to compare," and the term *para + deiknunai* meaning "to show." Paradigms, then, are patterns of thought that are destined to be compared and contrasted with each other. In the West we have taken this comparison to extremes. Contrasting ideas are often presented in rigid opposition—in binary, dichotomous terms that are black and white, like "saint and sinner," "church and state," or "scientific and unscientific." The contrast often results in conflict because we are predisposed to taking an adversarial position, one of showing that an opponent is wrong in order to confront the opposing position.[2]

As a consequence, we tend to think that change always involves opposition. We speak of a new paradigm versus an old paradigm, assuming that the one will defeat the other. But this is not really the way paradigm shifts occur, according to Thomas Kuhn, author of *The Structure of Scientific Revolutions* and the man most associated with the term. So what exactly do we mean when we use the word paradigm in this book?

Let's go back to Kuhn, who did not coin the word, but gave it a specific meaning in his influential book, a meaning that emphasized both the "entire constellation of beliefs, values, and techniques" shared by the members of a particular scientific community, and also the implicit assumptions and intellectual structure upon which research and development in a scientific field of inquiry is based.[3] In short, a paradigm serves both to bond a community and to substitute for explicit rules of methodology. Thus, the scope and interpretation of research is contained within the proverbial box of conventional perception. Results that fall outside the box are normally unseen or discarded unless or until they are compelling enough to persist in consciousness, which can then cause a crisis in the field. In a crisis, the unseen rules come out of the shadows and are not only seen again—they come into question. At this point, the possibility of a new way of seeing arises.

If a new paradigm is born, it is a fundamental reorganization of the tenets of the old. It is not cumulative, meaning paradigms do not come into existence through a simple amassing of information. A new paradigm includes elements of the old, but the restructuring of these elements is so different that it seems brand new. It is often perceived, at least initially, as more elegant, simpler, or closer to the truth.[4] I have found Kuhn's operational definition of paradigms to be useful, and in subsequent chapters, I propose what could be called a paradigm shift in terms of the reorganization of disciplinary knowledge and a reformulation of how a community of faculty might interact. The way Kuhn uses paradigm choice and the way I am using original thinking has some elements in common. At the same time, I feel a need for a broader working definition of the term paradigm, one that derives its origins from nature, as I explain next.

The Patterns within Nature

Any particular paradigm exists within a larger sociological field—a Zeitgeist or spirit of the Age—that influences it. But even this Zeitgeist is not the ultimate source, for it, too, is derived from a larger movement of consciousness. The ultimate source of paradigms is the patterns within nature.

The idea that Nature provides the metapatterns in which biological, sociological, and psychological patterns unfold is controversial but not unheard of. The biologist Rupert Sheldrake has basically said the same thing, calling these patterns "morphic fields"—or nonmaterial "patterns of influence."[5]

Morphic fields work through a process of "morphic resonance" in which causal influences are transmitted in a nonlinear way through space and time by matching vibratory patterns. The continuing presence of vibration is what enables the past to bleed into the present. These patterns, according to Sheldrake, are cumulative. They tend to repeat and form a web of habits within a particular field or natural system. They appear eternal only because they have become so deeply habitual. In actuality, they are evolutionary, according to Sheldrake. They evolve and change over time.

I wholeheartedly agree with Sheldrake that the laws of nature (perhaps better thought of as the habits of nature) evolve over time. I also agree that vibrations of the past continue into the present and said so earlier; the coexistence of the past and present is an important aspect of what I mean by original. But the coexistence of the past and present is not the same thing as simple accumulation from past to present. This way of thinking is too linear. Something else happens in a living cosmos, which is that energy is being continually reorganized, sometimes radically.

Sheldrake realized this and allowed for both gradual and sudden changes in morphic fields, parting ways with Darwin, who emphasized gradual change through natural selection. But people reading Sheldrake through the lens of a linear progressive paradigm may misconstrue his meaning as wholly linear because of his emphasis on accumulation.

The way I see it, the organization of thought into patterns (paradigms) is part and parcel of the organization of the larger patterns of nature. Moreover, these patterns of thought and nature are at their core, cyclical; they not only accumulate, but they also dissipate and reconstitute (renew) themselves at the appropriate time. I further address the cyclical nature of paradigms in a moment, but first, I want to discuss the nature of thought itself and its role in creation.

What Is the Nature of Thought Itself?

Of all the energies in the cosmos, thought may be the most subtle but also the most powerful. Thought, I believe, is the force that moves matter into manifestation, from the implicate to explicate order, much like the elusive Higgs boson scientists have been searching for. In short, thought creates form.

Thought, I propose, is the animating principle of the cosmos—what the ancients called "Fohat,"[6] or what Grandfather Leon called "the fifth element of time." If this is so, then thought is the primary driver of formative causation.[7] This is why James Jeans said, "The universe begins to look more like a great thought than a great machine." Without thought, nothing would happen. With thought, all is possible.

Thought is a particular form of living, subtle energy. Like all energy, it cannot be created or destroyed, but it does reorganize. It can gain strength through repetition, but it can also dissipate through disuse or undergo transformation when it reaches a certain threshold. Thought is nonlinear, and in all nonlinear systems, repeated iterations may cause powerful and unpredictable effects.[8] The dynamics in the "thought system"—by which I mean the patterns of thought that compose a paradigm—may become disturbed and unstable, resulting in dramatic and rapid change. This can be dangerous, but it also can be an opportunity for substantive change, one that breaks out of the constraints of the previously prevailing paradigm. It is then that an original thought pattern or new paradigm can be born out of the ashes of the old.

This original paradigm is not new in terms of a progression of discrete events. To the contrary, the new paradigm is only new in the sense of renewal. In short, paradigms conform to the cycles of nature. They go through the same cycle of birth, growth, maturation, decline, death, and rebirth (or renewal). To think of paradigms in this way requires no less than a shift from inert ideas to living thought, and *from line to circle.*

The Cyclical Nature of Paradigms

Everything I have said so far has been a prelude to repositioning the concept of paradigms in the context of nature's overall cycles of renewal. Paradigms are a pattern or constellation of thought that originates in (and is derived from) nature. As such, paradigms must be cyclical because nature is cyclical.

I see the total volume of thought as a grand movement, analogous to a hydrological cycle of ideas. Ideas, like water, filter through all levels of society. They circulate and coalesce much like water does, from individual drops of thought to rivers and oceans of thought. They pervade not only scientific communities, but much of society, like the oceans pervade much of Earth.

Within the human community, there is a biological and sociological field pertaining to every level through which thought filters down—from the planetary Zeitgeist to the nation-state Zeitgeist, to subcultures of all kinds, including structures of scientific communities that formulate theories that then feed back into the greater society. It is not only scientists and scientific theories that affect the Zeitgeist or whole of consciousness.

The most widespread ideas become tacitly accepted—which is to say nobody questions them, or even thinks about them anymore. Eventually, however, some thoughts lose their charge and dissipate, evaporating into the atmosphere. These unoriginal ideas rise to the skies, and in that creatively charged atmosphere, we might imagine them in conference with other "idea clouds," hobnobbing, merging, and reformulating into a greater and more original cloud. They are then ignited into action by a lightning bolt, eventually returning like fresh rainwater to earth, wherein they begin the cycle anew.

But what about ideas that do not circulate as freely? For whatever reason, some schools of thought are hidden and not spoken about publicly during a particular age. For reasons of safety, or simply because their way of thinking is so different from that of the mainstream, these ideas go underground. But this, too, is analogous to the water that goes underground and refills aquifers. Even while below the surface, the underground waters retain a relationship with the above-ground waters and the waters held in the sky, as the Hopi tell us.[9] And, like subterranean water, even esoteric or secret ideas resurface when it is time, bubbling up from underground springs.

I believe Kuhn was correct when he said paradigm shifts do not derive from a mere accumulation of ideas, but there is a relationship between old and new ideas that Kuhn mostly ignored. So-called new ideas are not only reformulations, but also a recycling of the old. And the older the source, the newer the ideas seem. Kuhn was on the precipice of exploring this when he noted that, in certain ways, Einstein had more in common with Aristotle than with Newton. He concluded that paradigm changes can be nonlinear, which is certainly correct—but why not say they are cyclical?

Consciousness, it seems to me, must unfold as all of nature does—in cycles. Paradigm shifts are but manifestations of this same movement.

Is it really so unusual to imagine our thinking as akin to the way water circulates? After all, we are composed mostly of water. Could it be that water thinks too? I say yes. Ultimately, we think not only as water does, but as all the elements do—for the elements do not act independently; they act in concert to create the circle of life. It is the unique combination of the elements—the way they move and interpenetrate each other in a ceaseless flux—that forms the matrix of existence.

I propose the following definition of paradigm, which came to me intuitively, but seems to capture the overall movement of thought in the way I am intending:

> A paradigm is a constructed view of the world that originates in the patterns of nature and pervades societies, subcultures, and individual sentient beings in cycles of unfolding consciousness in the same manner that nature unfolds; each emergent idea includes what has already unfolded and is now tacitly understood. The entire movement of consciousness is contained within the Circle of Life and the cycle of the elements—from the creative fire of imagination fueled by air, cooled by water, and solidifying into earth—until such time as earth dissolves into water, water becomes air, air becomes fire, and fire spirit, initiating the next grand cycle of the process and giving rise to everlasting cycles of renewal.

Since this definition came to me intuitively, I prefer not to analyze it, but instead to approach it through feeling. The overall feeling I have is that our thinking has a lot to do with what the Earth is thinking—that our thoughts are beholden to a larger sense of timing that is unbeknownst to us, and that this timing is circular. There is a sense of enfolding and unfolding that is not headed in a single or linear direction, but cyclically recurs, ever rebalancing over time.

I feel the nature of the elements and thought as one interconnected whole. I sense that thought need not separate us from nature, but can connect us to nature through integral and conscious participation in the circle of life.

Although intuitive, this definition is not out of alignment with what I know to be true about paradigms. It does not oppose the concept that paradigms are constructed views that privilege certain patterns of thought while inhibiting others, but it places the origin of these patterns in nature herself, which is where it should be. It also does not deny that ideas often rely upon previous assumptions that can be invisible to the thinker.

This definition does, however, switch the emphasis—and the origin of paradigms—from the individual thinker to society and, ultimately, to nature. In so doing, it emphasizes that cycles of transformation occur in keeping with the flux and interconnection of nature, rather than according to rationally constructed human thought.

This definition resonates with me because it seems to poetically express much of what I have been writing about previously in more prosaic language, but if the definition does not yet resonate with you, dear reader, please be patient, as the following pages continue to describe the movement of thought.

You Say You Want a Revolution? Well …

We tend to connect paradigm shifts with scientific revolutions, but we invariably miss an important clue to the real nature of that relationship. The Old French "revolution" is derived from the observation of celestial bodies that revolved and is originally from the Latin *revolutus* or *revolvere*—meaning to "turn or roll back." This is why social revolutions were once understood as cyclical; they seemed to signify a return to power of that which had previously fallen out of favor.

When particular ideas or particular classes of people have been suppressed for too long, they eventually rise up. Whether a revolution is peaceful or violent depends on how the resurgence is received by whoever is in charge, and upon the attitude, patience, and wisdom of the protestors. Peaceful revolutions, like India's movement for independence from Britain or the "Idle No More" campaign begun in 2012 by Indigenous peoples of Canada, are mostly guided by resurgent wisdom. The same could be said about the various ways in which women are returning to leadership all over the world.

The cultural dynamics of political revolution are complicated and beyond the scope of this book, but the point I wish to make is that particular ways of thinking have their own timing, embedded as they are in the larger movements of energy in nature. When one aspect of thinking is highlighted, another falls into shadow, but eventually that shadow returns to the light and pushes what was in light back into shadow. The turning of the wheel often feels excruciatingly slow to humans, which is why Grandfather Leon wisely counseled patience. He said, "In time, a negative becomes a positive and a positive becomes a negative." It is not so much about what is correct as it is about what is receding or now emerging.

The same can be said about the way we think about thinking itself. For centuries we believed that our thoughts came from nature—and then we did not. Now, we are again beginning to ask questions. Are our thoughts really generated intrapsychically, by ourselves alone, independent from nature? Or are our thoughts a reflection of nature's larger cycles of unfolding, as we once believed? Which is correct, or perhaps more importantly, which is now on the rise?

Are We the Receivers or Generators of Our Thoughts?

As we explored in Origin, we first came to see ourselves as the generators of thought during the European Renaissance and in post-Renaissance thinking. This way of thinking shifted our concepts of time, space, progress, consciousness, rational thought, and what it means to be human. All of the above were once defined by relationships between humans and nature, but now are increasingly defined by separation.

Mind was separated from nature for the purpose of reaching a more objective understanding of nature. Unfortunately, it has gone too far, which is why it needs correction. This is not a correction as in right versus wrong; it is a course correction toward greater balance and wholeness. But the question under consideration: "Are we the receivers or generators of our thoughts?"—is not simple. Little Bear has asked that exact question, and it makes for a good dialogue question because it does not have an easy answer.

If the origin of paradigms is, in fact, nature and society, and thought transfers from society to the individual, then what is the role of the individual thinker? If the thinker is only a receiver, how is it possible to transfer thoughts from one individual to another? How exactly does thought move and spread?

The key to this apparent paradox lies in the nature of thought itself. According to Bohm and Krishnamurti (and most recently, Malcolm Gladwell) *thought acts like a virus*, and although they do not make a point of saying it, viruses are alive. Viruses spread through the same mechanism as cells use to communicate with each other, only viruses spread much faster than regular cell-to-cell movement. Thought spreads in a similarly rapid fashion. Today, thought is arguably dispersed faster than it ever has been before through the Internet and social media. This can be good or bad (coherent or incoherent, true or false)—but in any case, thought spreads.

Thought can move suddenly, as in a downpour of inspiration in a brainstorming session. But thought can also move like puffy, cumulous clouds, steadily, if at times imperceptibly. This is partly why thought is often associated with clouds and is even depicted in cartoons as a cloud hovering above a person. A neutral or pleasant thought is drawn as a white cloud, but if a person is angry, the artist typically draws a black cloud. These seemingly fanciful cartoons are a clue to the real existence of thought and crudely mimic the way thought operates.[10]

In ancient esoteric traditions, or modern reformulations of these under the name of "theosophy" (literally "divine wisdom" or a synthesis of science, philosophy, and religion), thought vibrations are considered to be real—as real as light, sound, and electricity waves. These vibrations can be seen by clairvoyants or others who have the capacity to perceive the subtle realms.

The visual aspect of thought is called a "thought form." Thought forms are energy constellations similar to the ones that cartoonists depict, but they also commonly merge into the overall aura or egg-shaped cloud that envelops a person. Thoughts take on particular colors. Brown or brownish red is associated with instinctual impulses; various tones of red with sensuality or love (including a beautiful rose red with unselfish and devoted love); yellow with higher knowledge or intellect; blue with religious devotion and compassion; green with spiritual intelligence and ability to navigate flux and change; and various merging of colors for every other intermediate state of mind.[11]

Typically, people identify with their thoughts as theirs alone simply because they have become habituated to them. But the thought forms that surround a person are not truly theirs because they originated somewhere else. A person may augment or add intensity to a given thought, and depending upon their level of awareness and the strength of their personal power (in Sanskrit: *siddhis*), can send it on to another with a specific intention. The sending of thoughts with ill intent is a form of black magic; the sending of thoughts with positive intent is a form of prayer.

Thus, the individual is a generator of thought only in the sense of what they do with it, which is to say that we are more of a transformer of thought than an original power source. People have a responsibility in how they receive thought and what they do with it, but they are not the sole creator of thought.

The primary mode of being human is receptive. Our bodies are vessels that receive light energy from the Sun, the breath of air from the plants and trees, liquid nourishment from the clouds, and oils and minerals from the earth. Our inner existence depends upon receiving energies from without.

The same is true with regard to thought: we are primarily receivers. If we want to expand our creativity, we have to have the courage to let go of what we are presently thinking and let in new thoughts. As long as we are preoccupied with thoughts that came before, those thoughts remain around us, and we cannot receive new thoughts to nourish us. And if we chew on thoughts like a cow chewing her cud, we eventually become habituated and stuck with those thoughts. In that sense, the thoughts become ours because they form our character. Once we become associated with certain ways of thinking, people expect us not to deviate from them, and we, in turn, become afraid to step out of our comfort zones. We act in predictable ways and become a creature of habit. If we do not realize this pattern soon enough, it becomes almost impossible to change.

The act of holding onto thought is, in effect, physically unhealthy. The release of thought, on the other hand, is liberating, which is why enlightenment is spoken of in such terms. *Moksha* in Sanskrit means "liberation." We are liberated from our past karma by releasing the attachments we have to our various thoughts. The extent to which we learn to freely release our thoughts determines the ease with which we receive fresh (original) thoughts. Original thoughts do not have attachments; they are not yet "tainted"—as Lloyd Pinkham of the River People put it during the 1999 SEED dialogue.

This need to release thought pertains even to the thoughts we identify as knowledge because, ultimately it is unhealthy to hold onto knowledge as well. Knowledge is available to us and will always be available to us, if we are open to Spirit. We only need to go to "the natures"[12] and ask with an open heart at the appropriate moment. The releasing of knowledge is therefore healthy, as Socrates understood and practiced throughout his life.

It is important to respect the inherent nature of thought, which is to move. We can then recognize the influx of appropriate knowledge as it arrives, and let it go when it is no longer needed. When we let go of thoughts, we clear the space for other, more original thoughts to enter. In fact, the less attached we are to any specific thought, the more we clear the way for original thought to find its way to us. We do not build academic knowledge or erudition in this way, but we do build the capacity to become wise. We become "spiritually intelligent"—which is to say we become intelligent on a moment-to-moment basis.

In sum, thoughts and thought forms have a peculiar life of their own and are ultimately not the property of any one person. It bears repeating that we still say in English: "A thought *came* to me"; we do this because we intuit that the thought is not something we own, but something that comes into our field of awareness from someplace else.

Because thought is not the property of any one person, thoughts can be shared. Thoughts are shared in a manner similar to how sound vibrations are shared—through a matching of frequencies. People in the same room often share the same thought because the thought has a physical presence. This is also, by the way, why many of us forget what we were thinking when we get up and move from room to room. All it means is that we forgot to bring the thought with us. If we return to the other room, we find the thought is still there, and we suddenly remember it!

Thoughts have a local presence, but thought is a subtler vibration than sound, and therefore can travel much faster. Sound can only travel at about 750 miles an hour (in air), but thought is believed by some to travel even faster than light, essentially instantly, over any distance on the globe between like-minded individuals. Why is that? The power of the connection between people of like mind apparently draws the thought between them, but the power is also in thought itself, which seems to have magnetic properties like electricity.

The magnetic quality of thought is at the root of much of the ancient esoteric sciences of mental attraction, sometimes repackaged as "the power of positive thinking," "The Secret," and various other pop psychologies. These pop psychologies are half-truths; they play with powerful principles not easily understood and too often mislead the public. Thought may have the power to attract, but that does not mean it should be used to get things, which is just selfishly augmenting our already oversized egos. Thought is a doorway to a greater power that lies in nature, and we are only a channel. We can learn from this, but in my view, we should always respect nature as the original teacher.

Chapter 11

To Make Thought Whole Again

What Is the Process and Function of Thought?

Thinking about the process of thinking is necessary because, contrary to popular belief, thought is real in itself and has real consequences. By this, I do not only mean that thoughts create internal changes within the thinker that can be measured by an electroencephalogram or EEG, although this is true. It is more that these internal changes lead to further changes, sometimes profound, in the outer world. Look around you. Much of what we take for granted, be it a road, a building, a city, or, for that matter, a bird's nest, was first conceived in thought. Thoughts play an integral role in perception, motivation, *and action.*[1]

Let's examine the subject a little closer. In common usage, thought can refer to both the act of thinking and its resulting ideas. We, however, tend to focus almost exclusively on the content of thought—the end product—and ignore the process of thought. It is not just the end result of thought that is important. How thought originates; how it functions; how it affects our consciousness; how it spreads like a virus to infect all of society and nature and how nature and society affects our individual consciousness—all of these aspects are integral to the whole. They are part of a seamless continuum—what Bohm called "a single unbroken flow."[2] The process cannot be rightly separated from the content of thought, which is merely the end result.

I am saying that thought must be regarded as a whole, inclusive of both process and content. One cannot be considered without the other. When we attend only to the content of our thoughts we fragment that whole. The fact that we do this habitually, that we mistake the part for the whole, is the reason why I was moved to write this book. We must *make thought whole again* by attending to the process of our thinking.

The importance of attending to the process of thought was succinctly addressed by Bohm in a passage from the chapter "The Art of Perceiving Movement" in his book *On Creativity*. He urged us to "attend to" and

> be aware of thought as such; that is, as a real movement that is actually going on, inwardly and outwardly, with real effects of very widespread and deep significance that interpenetrate and ultimately merge with the whole of the reality in which we live. We cannot do this if we go on with the prevailing tendency to concentrate almost exclusively on the problems which are the results of our thinking process, and thus fragment these results from their origin in the thought that automatically and habitually "copies" inappropriate patterns. If we fail in this way to give proper attention to the origin of our problems, then the very thought that is aimed at solving them will be of the same fragmentary and confused nature as that which is producing them so that what we do will tend to make things worse, rather than better.[3]

These important observations by Bohm are a small window into a lifetime of work examining the relationship of thought and consciousness to the whole of reality. Bohm was painfully aware that the effects of fragmentary thinking affected more than the individual. With perspicacious insight, he pointed out that it was "fragmented thinking (that) causes a fragmented society." He also carefully distinguished between "fragments" and "parts," noting that a part is intrinsically related to the whole, but a fragment is not.[4] The word "fragment," he pointed out, is related to the word "fragile" and is equivalent to the breaking of a watch by using a hammer as opposed to the careful dismantling of the watch into parts that could be put back together again.[5] Once we have fragmented consciousness, we can never reconstruct the whole, just as "all the king's horses and all the King's men could never put Humpty Dumpty together again."

Fragmentation is an inculcated attitude that "disposes the mind to regard divisions between things as absolute and final," Bohm explained, "rather than as ways of thinking that have only a relative and limited range of usefulness and validity."[6] If this were an occasional lapse of mind, that would be one thing, but it is not; it is the normal way we perceive. Our way of thinking regularly produces disjointed, irrelevant fragments of ideas rather than rich, contextual wholes. This is the legacy of perspective. We persist in seeing reality as composed of isolated things in themselves rather than as parts in relationship with the whole.

Is All Thought Inherently Divisive?

Perhaps the problem lies with the very nature of thought itself. Krishnamurti provocatively stated that "thought in its very nature is divisive."[7] Why did he say this? Let's first try to understand what he meant before moving to question the premise itself.

Krishnamurti, like Bohm, saw thought as a movement of consciousness. This movement cultivates desire, which leads to the idea of accumulating more, which implicitly separates one from another. Krishnamurti said, "The very nature of accumulation brings about a division between you and me."[8]

Earlier, we traced the division in human consciousness to the advent of linear perspective. Although it is true that perspective divides the world from the seer, and this became known as the realistic way to see the world (at least in the West), we cannot simply blame this tendency toward division on perspective alone because it does not answer the question of what brought it about in the first place. Krishnamurti is zeroing in on the root cause.

What is it in human nature that makes us imagine that we are separate from nature and from each other? Krishnamurti says it comes through desire. This makes sense to me. Desire is one of the most deep-seated human emotions. Religious texts speak about desire as an inherent attribute of the Universe; it is what makes Creator create. In the Rig Veda, desire is the primal seed of Spirit. Desire creates movement. Out of nothing, comes something. Something stirs. The seed that has never seen the light grows toward the surface. What moves it to become? Desire.

But Krishnamurti goes further; he specifically equates desire with the desire for more, the desire to accumulate. What comes to mind is the beginning of the Tao Te Ching:

> The Tao that can be told is not the eternal Tao
> The name that can be named is not the eternal name
> The nameless is the beginning of Heaven and Earth
> The named is the mother of the Ten Thousand Things
> Ever desireless, one can see the mystery
> Ever desiring, one can see the manifestations.[9]

It seems that desire creates manifestations, which appear in many forms. These appearances are deceptive because they are often confused for the original, eternal One, which they are not. They are representations of the One; they represent the One in abstract form, and there are an infinite variety of manifestations depending upon the imagination of the mind.

Now, there may be a desire to accumulate which is not abstract, that is wholly practical, such as storing food for the winter to survive until spring. But the outward practice of accumulation inevitably spills over into the psychological realm. It is here where the seed of fragmentation occurs which, according to Krishnamurti, is psychologically harmful.

This fragmenting of consciousness through the desire for more underpins the concepts of time, progress, identity of self as a separate individual, accumulation of knowledge, passing on of that knowledge through education, and so forth. In each case, the desire for more is what propels the individual toward alienation from the whole.

Krishnamurti equates thought with movement, but also thought with time itself, saying that "Wherever there is [psychological] accumulation there is time."[10] This is an important observation because it reverses the usual emphasis on time as a given, and accumulation as something you do over time.[11] It acknowledges time as a human construct created by the movement of thought. We saw how Little Bear equated psychological identity with time when he stated "I = Time"; Krishnamurti is doing something similar, but perhaps emphasizing the cumulative aspect of thinking even more so.

The equating of the accumulation of knowledge with psychological time is as good an explanation as I have heard regarding what brought about the shift to linear perspective because, before we could invent perspective, we had to store up man-made knowledge. It was the storing of knowledge that led us to imagine that we could influence the future we now saw as being ahead of us.

The idea of storing knowledge may seem like a given—as obvious as the concept of time. But both are human-made constructs. We used to trust our intuitive abilities to receive knowledge from nature on an as-needed basis. We can still do this today. We do not have to give up the knowledge we have gained as a society, but we do need to set it aside—psychologically—if we are to open ourselves up to a larger truth of how to live in the moment that is not contained in any book or manual—or even in any guru's teaching.

Everything Krishnamurti is unveiling as illusory: time, progress, the accumulation of knowledge, the passing on of that knowledge through textbooks and authority figure teachers—is fundamental to the Zeitgiest of our age. It is part of the invisible tacit infrastructure, the shared social conditioning that underlies how our modern society has been operating for some time.

This Zeitgiest—this inexorable movement of energy that pervades and affects the way we perceive, think, and act—is difficult to see precisely because it is all-encompassing. This is why a statement such as "all thought is divisive" is useful. It is actually a koan; it ties the mind up in knots to think of thought as inherently divisive. It makes us stop thinking. And when we stop thinking, stop searching, stop asking questions, and are just silent, empty—then there is the possibility of original thought arising.

A thought that arises in this manner is *not* divisive, which is why Krishnamurti also said: "If there is an awareness of the beginning of thought, then there is no contradiction in thought."[12] If we simply observe our thoughts without attempting to control them, thoughts flow as nature intended. The attempt to control our thoughts and forge our own path is what separates thought from its origins. The kind of thinking that seeks to control itself is egoic and divisive; thinking that arises effortlessly, without desire to control, is not.

Krishnamurti's insights into individual thought processes are equally relevant when we consider the larger movement of collective Western thought. Our culture as a whole has basically embarked upon the hero's journey as written about by Joseph Campbell.[13] The hero's journey is a way of discovery through departure away from one's homeland. It is symbolically equivalent to a necessary separation from Mother's womb and is mostly associated with a masculine path of initiation and outward exploration. Our entire society, men and women both, have gone in this masculine direction of movement away from origin—but this is not the end of the story. As I keep reiterating, any extreme path bears the seeds of its own destruction. The hero must eventually return to his homeland.

The Return

The pathway of Return begins when the masculine journey away from origin has run its course. The urge to return back to origin is distinctly feminine in nature. The Return leads back to Oneness, wholeness, and healing. It, too, will be undertaken by the whole society, both women and men. And it is already underway.

> *Returning is the motion of the Tao*
> *Yielding is the way of the Tao*
> *The ten thousand things are born of being*
> *Being is born of not being.*[14]

There is good reason to believe that the outward arc of Departure from our origins is coming to a close, and the inward arc of Return is commencing. We are in the early stages, to be sure. But we are starting to question, starting to let go of at least some of the assumptions that have constellated the modern psyche—what Krishnamurti correctly saw as our social conditioning around time, knowledge, and accumulation.

We must continue to question our conditioning if we are to escape from the supposed inevitability of linear progress—a pathway that has short-term benefits for humanity, but puts the whole Earth community, including ourselves, in jeopardy for the long term. We must awaken to the simple truth that our existence is bound up with that of other creatures; that all of existence has the "Right to Be," as Moonwalker and O'Brien-Levin put it.[15] In realizing that human-centered thinking is not a sustainable path, we are returning to the essence of what we are and what we always have been—beings who are of the interconnected Earth, not transcendent from our origins.

Ecotheologian Thomas Berry called this direction "The Great Work," a reawakening of the intimacy between human-earth relations, ushering in the "Ecological Age."[16] David Korten aptly adjusts this to: "The Great Turning,"[17] a phrase that Buddhist ecophilosopher Joanna Macy has done much to popularize.

Macy sees three great revolutions in human history: the agricultural revolution; the industrial revolution; and the revolution we are now entering. The industrial revolution has devolved into a "militarized industrial growth society" which is life-destroying, and this degradation is leading to an uprising from the people, now underway. It is a movement that is, in her words, "life-sustaining" and aware of "our relation to each other and to the sacred living body of Earth."[18]

The first principle of this life-affirming revolution, according to Macy, is gratitude. I whole-heartedly agree with her: beginning in gratitude makes all the difference. If we once again recognize, as Indigenous peoples have for millennia, that life is a circle and therefore whole, we pray to Creator out of gratitude for all the blessings Creator provides us right now; we do not pray for something in the future to complete us. We have so much to be grateful for, beginning with the Sun that gives us life, warmth, and energy; the clouds that bring the rains that nourish the trees and plants that give us and other animals the gift of breath that we give back to them in a sacred circle; the majestic mountains, thundering waterfalls, magical rainbows, inspiring birdsongs, and the overall totality of awe-inspiring beauty and diversity on this planet. The world is whole and complete, and we are a part of this world and therefore whole and complete. If we begin from wholeness we remain in wholeness.

> *Aum.*
> *That is the whole*
> *This is the Whole*
> *From Wholeness emerges Wholeness*
> *Wholeness coming from Wholeness,*
> *Wholeness still remains.*[19]

Remember this if you do not remember anything else from this book: The origin of thinking is thanking. Original thinking is thinking that begins in wholeness, in gratitude for all there is. Thinking that begins in wholeness remains in wholeness, even if it employs abstract thought for limited purposes. We are always in relationship with the whole.

Chapter 12

To Think without Separation

This early morning I am inside my living room seated at an old drafting table that serves as my writing desk, dreamily gazing upon my familiar Sandia Mountains. It is a bitterly cold morning, and snow is gently falling. The steep and rocky edifice of the mountain is white only in the places where the snow manages to cling before falling to earth, making the mountains appear as if they were dusted with powdered sugar.

Next to my window the snow has managed to accumulate two or three inches, a not insignificant amount for this neck of the woods. I get a sensation that the earth is resting, cozying up with its newfound blanket of snow. But if the ground is sleeping in, the finches and jays are wide awake and hungry, quickly discovering our "restaurant," as my wife and I like to call our feeder. Then, just as suddenly they are gone, tipped off by a scout atop a nearby piñon tree. It is then that the most amazing thing happens. A golden eagle majestically lands and then perches on a thin wrought iron pole that supports a set of wind chimes. Its landing rocks the pole, which rocks the chimes, and it clearly enjoys the sensation. This astonishing creature is only twenty feet in front of me, and I do my best to inhabit his mind, seeing through his eyes: bobbing of head … right, left, up, down, surveying everywhere, playfully attentive and aware … listening, watching, ready. The presence of the eagle inspires me, but not because of any particular quality or set of qualities—not his wingspan, or his eyes, the white coloration on its tail, or any one thing. It is the whole of eagleness that is beautiful and awe-inspiring even as I do not know why.

I know that my humanness is somehow completed through engagement with the eagle, and I know that I feel less complete when I disengage to mull over routine thoughts. The alertness of a wild animal alerts me; I know it is a special opportunity. The other critter is different from me, and it is the difference of two presences, two distinct essences, that inspires learning. I become wiser through the interaction, and this is likely because wisdom and inspiration require the influx of spirit. I do not become inspired all by myself, and I do not really think by myself either. My thoughts are ignited when another essence sparks mine; awareness is born not between my ears but in and through encountering another.

In short, the origin of thinking is relationship. We share a field of awareness with an aware cosmos, and it is out of that field that individual thoughts arise within us. These thoughts come first to our hearts, and strike a resonance, a pulse that matches the pulse of nature. Hearts think in that way. My friend Black Eagle Sun says there is no difference between the heart of nature and your own heart, and in a sense that is true because they are connected to the same vibrational source. What we call thought originates in the pulse of nature and the pulse of our own heart; it then travels from our heart to our brains, later to become words through the filter of language. Both our brains and our languages are filters. Thought originates in the heart of nature, in and through relationship. We do not initiate thoughts in the way that we imagine, nor is thinking unique to humans, even as "thinking about thinking" might be.

What happens when we do not place the origin of thought in nature? We fall into the trap of seeing ourselves as the primary source of energy, light, and thought—we believe that we are our own sun. Moreover, once we imagine that thought arises within ourselves alone, it is a short step to denying that consciousness exists in nature at all. We have rejected the *anima mundi* (or "soul of the world"), a concept we believed in for centuries, in order to reimagine that only humans are endowed with soul.

In the same way, we have come to think that only humans are conscious and therefore the pinnacle of evolution. We have made ourselves into a kind of demi-god. This has been our pathway, a road that has given us a massive superiority complex in total opposition to our forebears, the ancient Greeks, who counseled against such hubris. Not only did we ignore their advice, but we also convinced ourselves that we were entitled to dominion over the rest of creation.

Dominion over nature may seem a Christian concept, but it really is not any more. It now encompasses all of Western society, as noted by Thomas Berry and others.[1] It pervades our legal system, according all rights to humanity and none to other life forms; it is in our universities, where nature is divided up into separate subjects according to their particular human applications. It is most obviously in our corporations, some of which pollute wantonly unless or until they are reined in by government regulation—but it is also in our governmental structures, designed largely if not wholly for and by humans. It is in our national parks, for instance, which are not so much a sanctuary for animal life as enlarged zoos to amuse humans.

The entrance at Yellowstone says: "For the benefit and enjoyment of the people." The Western concept of conservation is lacking, or not whole, as Leopold pointed out, and this is probably because it comes from a sense of lack, a sense of scarcity. Little Bear once passed through Yellowstone, to him still part of Blackfoot territory, and asked one of the rangers: "Is the park a model for how you would like the rest of the natural world to be kept, or is the park a protected area so that the rest of the land can be destroyed?" Not surprisingly, he received no answer.

Some may think it obvious that humans are superior, given our prowess in technological invention. We have become so clever that we are able to store fire at the end of a stick—what we call a "match." My friend Larry Littlebird, a Pueblo Indian from the Laguna/Santo Domingo tribes in New Mexico, finds this most remarkable, and I agree with Larry; it is. We have also created architectural and artistic wonders such as the Sistine Chapel, the Great Pyramids of Egypt and Mesoamerica, the supercomputer that calculated a pathway to the moon, and the smartphone with the same power in the palm of our hand. Of course, we still have not developed anything as strong and flexible as spider silk or a ceramic as durable as an abalone shell—but put aside this debate and assume that humans are superior, at least in certain ways.[2] Then what? Does that mean we should only look after ourselves?

At best, modernity is a double-edged sword. Our dependence upon rationality has enabled great technological invention, but it has also disabled our ability to see human beings as dependent upon the whole. Thomas Berry cogently wrote:

> The centering of value so extensively on the human distorts the place and role of the human in the structure and functioning of the universe. We fail to recognize that although the various components of the universe exist for each other, each exists primarily for the integrity of the universe. The human also, while noble in itself, exists for the integrity of the universe and for the Earth more than these exist for the human. Indeed, the human depends upon the larger universe for its existence, its functioning, and its fulfillment. Within the order of the universe the planet Earth provides the efficient, final, material, and formal causes that bring the human into being, support the human in being, and lead the human to fulfillment.[3]

The concept of the human as nested within the whole of nature was accepted throughout history, at least until medieval times, as Berry goes on to point out. Medieval theologians such as Thomas Aquinas, still emphasized the whole. Aquinas said, "The order of the universe is the ultimate and noblest perfection in things."[4] The wisdom of Aquinas, shared by Berry, may have gone out of favor for a time but, thankfully, its season is upon us again and will never completely disappear because it is perennial wisdom. Its return comes not a moment too soon; the necessity to reevaluate what it means to be human has never been more imperative.

To mainstream modernity, however, the role of humans is settled. We are the noblest perfection. This belief is predetermined once we equate progress with humans alone, and more narrowly, because we equate human progress with technological advancement. We have to stop making a one-to-one correlation between technological development and real advancement. The future is not just about what *things* will be there! It is also about what our consciousness will be.

I am troubled by the fact that many ecologists seem to accept sustainability as our ultimate goal. Can't we go beyond sustaining our present lifestyles and begin to rebalance, replenish, and restore the way of life we once enjoyed? I believe we can. I prefer these "re" words because they imply cycles, not lines of progress. Organic growth is great; everything must grow, but everything must also die and return to source to nourish others. We keep pushing against nature to do something that is ultimately impossible—growing and expanding forever. Nature is finite. So are we.

Now, I realize all too well how difficult it is to change our current concept of managed growth because growth itself, and particularly economic growth, has become an idol—a concept that few dare to challenge. We will shortly delve into the subject of economic growth in more detail because it is essential that we deconstruct its idolatry. But first let us establish the underlying reason for our current fixation on economic growth: the delusion that we are destined to progress forever away from origin while the rest of nature is somehow static. This same worldview that isolates human progress from the context of nature also insists that progress cannot be rolled back: "What's done is done," we say, even as there are many examples of things being *undone*, such as Ghandi's revitalization of cottage industries in Indian villages, or Cuba's revitalization of traditional farming during the oil embargo of the late twentieth century.[5]

We are repeatedly told that progress is unstoppable and must be accepted. My high school friend used to tell me, "You can't fight city hall."[6] Although that may seem like a reasoned position, it is merely an assumption, and one that enables the entrenched powers to control people by taking advantage of said acceptance. In my opinion, the attitude of the inevitability of progress is at the root of many of the ills of the modern world, including war, poverty, and many other injustices upon other humans and upon the entire natural world.

One of the most prescient thinkers in this regard was Bateson, who stated that human beings were doomed to total destruction unless we could change just *one* of the following three things:

1. The Western idea of humanity

2. The pace of technological growth and change

3. The pace of population growth

Of the three things, Bateson felt that the first—the Western idea of man—afforded the best possibility of change.[7] I could not agree more, which is why I was compelled to write on what it means to be human in order to reframe the Western idea of humanity in a way that would allow us and other species to survive. But to Bateson's list, I would add two others:

1. The Western idea of time

2. The Western idea of rational thought

I would add these and keep the same premise that any one of these things could change and we could survive, and here is why: The Western idea of time is so tied up with the Western idea of rational thought and the Western idea of man that they are inseparable.

At present, we believe that only humans can progress and that only humans can keep time. But what if we were to change our thinking so that we saw all progress as but an unfolding of what was already present—that, as Prechtel poetically said, "time is a plant grown from a seed"? We would then, by necessity, reorient our perception toward keeping pace with the unfolding of nature rather than trying to control her. And that would completely change our perception of humanity, which would then be seen as enmeshed in the whole, rather than separate, superior, and transcendent from nature.

Similarly, if we were to recover the original meaning of rational thought, embracing it again as a quest to discover the relationship of sacred proportion between things, enormous changes would occur. We would then deemphasize breaking nature down into separate subcomponents for analysis and reemphasize looking for the pattern that connects in keeping with the natural philosophers of old. Moreover, we would be reconnecting with nature, instead of only abstractly analyzing her. In either case, we would be reorienting our view of man, which would, as Bateson implied, change everything else.

In the end, it all comes down to how habitual thinking orders our reality. If we unmask our assumptions and become aware of what undergirds how we think, we change our view of reality. Right now, we believe that abstract thought is superior, and so we have designed all of our institutions and structures to match; our schools and universities, our professions, and industries, economies, our legal and medical systems, politics, and government are all based upon an imbalanced and undeserved reverence for rational analysis, separation into specializations, and polarized thinking. Debate (which separates), rather than dialogue (which has the potential to reconnect), is how we communicate. This must change.

The Problem of Abstraction in Academe

The current structure and form of higher education is overtly abstract and fragmented. As such, the university is the greatest obstacle to changing the way we think, and it also presents a great opportunity. If we can change the structure of universities by changing the way knowledge is organized and disseminated, we can change society for the better.

It is indisputable that universities hold a great responsibility in our society because they sanction what is (or is not) acceptable thought. They decide not only how we should think, but what subjects are worth thinking about. They delimit what fields or professions are permitted, decide what research is acceptable in any given field, and participate in determining the credentials for entering a particular field or profession.

Most importantly, it is universities that have standardized the practice of organizing knowledge into abstract academic disciplines. They have done this so completely and so successfully that its efficacy is no longer questioned, inside or outside academe. Instead, mainstream society has come to believe that the world really is divided up into subjects, which would be laughable if it did not have such devastating consequences. The unfortunate reality is that our current graduates enter their fields believing (in whole or in part) that the real world operates as it does in academe. It does not, of course. Real problems do not conform to disciplinary borders, much as we may wish them to.

Fragmented thinking is tacitly encouraged at the university through the division of subject matters into academic disciplines, each with their own building, their own culture, their own budget, personnel, and their own jargon. Our graduates then enter society as if they were a horse with blinders, unwilling or unable to consider what is outside their field of expertise. Outside academe, society as a whole has bought into the abstraction of nature; people really believe that we can understand nature by dividing it up.

Throughout my life, I have found the arbitrary division of academic subjects strange, and longed to change it. The fact that the organization of knowledge in modern academe is not as old as it pretends to be (it is only some 150 to 200 years old) should lend it to reformation.

The problem is, however, that the most successful disciplines are those that employ the highest levels of abstraction. Physics immediately comes to mind. The discipline emerges out of observation of nature; it is also an exacting discipline, seeking to portray the world in mathematical equations. To a large extent, this works fine, but it is nonetheless an abstraction, an overlay upon the physical land, much as in Abram's story of the King's map that opened this part of the book.

The success of physics has some deleterious effects, including the idea that Western science is the world's science; in other words, a superior way of knowing. The word science comes from the Latin *scientia*, which simply means "to know." There are many ways of knowing, but today, ways that are other than those of Western science tend to be marginalized or dismissed. They are labeled as "folk knowledge" or "myth," both of which are taken to be hearsay or untruth. I prefer the word "story" to "myth" because it is more neutral. A story could be true; a myth is assumed to be a falsehood.

Of course, science is not a monolithic block; the scientific community is made up of individual scientists, and certain scientists recognize the value of storytelling in keeping knowledge alive and vital. The mathematical cosmologist Brian Swimme is one such scientist who has embraced this tradition, bringing the art of storytelling to the entire arc of the cosmos. Swimme partnered with Thomas Berry, another original thinker, to pen *The Universe Story* and also wrote *The Universe Is a Green Dragon* and other books and documentary films that utilize the power of story, which is probably why they have been so successful.

Ultimately, the big bang is just another story, as Little Bear has said to me, not out of any disrespect, by the way, but out of respect for the keeping of stories. This reminds me of a story, a short but saucy one that the late Native American writer and activist, Paula Gunn Allen, told at the 2004 dialogues:

> Where I come from, God is a woman. So I can't deal with this Big Bang Theory. I think the Big Bang theory is a male mock-up. And I think it's cosmic orgasm and that's really cool. KABOOM! YEAH! WOW! That's great, but it couldn't be the beginning of anything; it just couldn't, because what man ever had one and didn't go back for more?

I have to agree with Paula, whose humor I dearly miss, even as I have the utmost respect for Swimme and other scientists such as José Lutzen-berger who have eloquently told the story of the universe from a Western scientific (and perhaps, male) perspective. The big bang is a story that satisfies our need for beginnings, and dramatic ones at that. Yet it may be just that: a story.

The reader may object, "But there is evidence for the beginning of the universe," and indeed there will always be if that is what you are looking for. Facts are made to fit theories, or as Goethe once said, "everything factual is already theory."[8] Now I must confess that I too have been pro-grammed to believe in beginnings because my first language is English. But I no longer believe in beginnings, or at least I no longer believe in one beginning because it makes so much more sense to me that creation occurs in waves, in cycles that do not begin or end, but recycle, perhaps in a spiral progressive pattern. Any one beginning is simply the start of a new cycle, just as December 22, 2012, was the start of the next 5,000-year Mayan long count, not the end of the world.

Physics is a most interesting discipline in that many of its theories and discoveries open a door to spiritual inquiry by whetting the appetite for what underlies reality. But physics does not ordinarily attempt to solve spiritual conundrums. In fact, the spiritual dimension is conspicuously absent from physics theory with some notable exceptions. Bohm's concept of "implicate order" is one case where something tantamount to God or Great Spirit is implied, because the implicate order is imagined to be the invisible source of creation that unfolds into the explicate or visibly manifest order.

But let us return to the success of physics as a discipline, a success that was envied by the rest of the academic world. This is why physics became the ideal upon which all the other sciences, including the social sciences, modeled themselves for much of the twentieth century. This proved to be a cardinal error for the social sciences, which kept trying to become something they are not and never will be: a hard science. Sciences such as psychology, sociology, anthropology, archaeology, and the like will never be able to predict the whole range of human behavior with any degree of accuracy because human thought and human society are too diverse and variable to be statistically forecast. The same goes for economics, which we get to shortly.

The social sciences, nonetheless, tried in vain to model themselves after the hard sciences. And ironically, they began doing this at a time when there was about to be a conceptual revolution in physics, even though physicists themselves did not anticipate this. Remember Lord Kelvin? At the end of the nineteenth century, he boasted that "There is nothing new to be discovered in physics now. All that remains is more and more precise measurement."[9] Then, in a few short years, everything in physics was turned upside down when relativity and quantum theory burst upon the scene. Physicists discovered, and still are discovering, a whole new way of conceiving reality that, I must note, is remarkably similar to ancient ways of knowing in terms of its emphasis on radical interconnection, process, and relationship. Fortunately, scientists like Einstein and Bohm were decidedly less arrogant than Kelvin. Bohm stated at the outset of his ambitious *Wholeness and the Implicate Order* that there will never be any "received and final notion concerning the nature of reality."[10]

One of the biggest ironies of quantum theory is that it is still not taught in school—at least not at the lower levels of education, or really even in college, at least when I attended a little more than a quarter century ago. Have we decided that quantum theory is just too revolutionary to be understood?

Thankfully, a different trend has occurred in the social sciences that is incorporating significant changes, particularly with respect to the role of the observer. Consider anthropology. When Bateson and Margaret Mead were research (and also marriage) partners in the mid-twentieth century, they initially conducted their scientific inquiries in the rather stiff but academically acceptable convention of the times. In short, they parked outside an Indigenous community and observed village life. By separating themselves from the subject they observed, they sought to maintain their scientific objectivity.

Bateson and Mead soon realized this did not work—that in order to really understand a village, you have to actually become part of it; otherwise, nobody will trust you. To their immense credit, they changed, and changed dramatically. They also began to follow real problems where they led, regardless of the boundaries defining any single discipline. In so doing, they became splendid examples of original thinkers who greatly expanded the scope of anthropology into multiple other fields, thus demonstrating how knowledge from one domain can cross-fertilize with another. Bateson, in particular, became associated with everything from psychology to systems theory, cybernetics, and biology; the same was true of Mead, if to a lesser extent. Together and apart, they helped change the world for the better.

Years later, qualitative research practices emerged within all the social sciences, formalizing what Mead and Bateson had already started. These practices are now termed "participatory action research" and include various other "participant observer" methodologies.[11] In fact, there are so many researchers living with Native communities nowadays, particularly in the Southwest, that Vine Deloria Jr. once deadpanned that a "typical Native family includes a mother, father, sons, daughters, grandparents, great-grandparents, and anthropologist."[12] All kidding aside, the trend toward qualitative research is a wonderful step in the right direction, not only because it includes the observer in the experiment, but also because it fosters research that is based in rich narrative rather than statistical analysis alone, often combining the two. I prefer a story to a chart or graph, but if you must show a chart or graph, tell me a story with it, please.

And with that, I will tell a story, one that may seem to be about a dry subject: economics. But I will make economics into a *wet* subject, reconnecting it to water and its other life-affirming roots.

Chapter 13

Rethinking the "Dismal Science"

What Are the Living Roots of Economics?

The seeming exception to the unpredictability that characterizes much of the social sciences is economics, which excels at predictive modeling. Economics, like physics, became successful by developing its discipline to the highest levels of abstraction. And although the ramifications of the discoveries made in physics are consequential—witness the creation and use of the atom bomb—the field of economics may hold even greater influence on the workings of the modern world. Villages, cities, counties, regions, and nation-states are all beholden to economists and economic models.

Economists shape the future of the entire planet in many unintended ways. Their main fallacy is to put the human first. They do this to such an extreme that today's economic policies have already been the undoing of countless species, and if not redirected, could be the undoing of our species. There is still time to change, but meaningful change will only occur if we radically rethink the purpose—and very soul—of economics. In short, we need to think originally about it, recovering the living roots of economics.

Centuries ago, economic thinking was in service to the common good. It was not dangerous or exploitative of Mother Earth because we had not yet separated our consciousness from her. Fast forward to the present, however, and that separation is near absolute. The perils of current economic practice are largely a by-product of modern abstract thinking that separates the human from nature. It is also a product of the separation of academe into distinct fields of research.

It was not until the late nineteenth century (1885 to be exact) that economics was established as a formal academic discipline.[1] The formation of the discipline forced economists to make certain choices to distinguish their field from other fields. At that time, academe was just beginning an amoeba-like division into more and more distinct disciplines.

Economics was one of the first social sciences, predating psychology, sociology, geography, and countless others. Because economics was one of the first social sciences, economists were motivated to make their field accountable, which is why they chose, like physics, to become a deductive discipline. In this way, the field could produce results suitable for statistical analysis, prediction, and control. This decision helped the nascent discipline achieve a measure of success and respect in the greater society, but it also meant the field had to jettison significant aspects of its former self. The latter part of the story is rarely told.

To establish itself as a deductive academic discipline, economics had to abstract a part from its former larger and more inclusive wholeness (from what is now known as "classical economics"). The new discipline is therefore sometimes referred to as "neoclassical economics" after Alfred Marshall, its founder.

Marshall was interested in the place of economics in history, and that is why he helped found the discipline. He was also sincerely interested in solving the social problems of humanity and stated as much. But Daly and Cobb make an important point, noting that Marshall "allowed his task to be determined for him by the discipline itself and not by the social problems."[2] In short, he furthered the discipline, but not so much the world. This is the unfortunate reality of academic life, in which disciplines become self-serving and do not contribute to understanding beyond their sphere. This is because disciplinary specialization and broader understanding are at cross-purposes. Understanding comes from rich context, and *disciplines eschew context,* choosing instead to wall themselves into their own ivory tower.

The early classical economists—that is, those economists who lived prior to the founding of the academic discipline—were a far cry from modern economists. Adam Smith, writing in the eighteenth century, more than a hundred years before the academic discipline was born, maintained strong historical, political, and humanistic contexts within his own philosophy. Smith understood that the economy was only part of the whole of human activity.[3] John Stuart Mill, writing about a half century after Smith, similarly understood that:

> Political economy is not a thing in itself, but a fragment of a greater whole; a branch of social philosophy so interlinked with all the other branches that its conclusions, even in its own peculiar province, are only true conditionally, subject to interference and counteraction from causes not directly within its scope.[4]

The above quotation was written in protest of a movement seeking to found a professorship at Oxford University a little over 150 years ago. The poet-philosopher Samuel Taylor Coleridge and other romanticists felt similarly to Stuart Mill, asserting the interdependence of economic issues with social, moral, and religious issues.[5]

But the historical roots of economics (the living roots that connected it to other philosophies and ways of thinking) were abandoned once the choice was made to form an academic discipline, and a deductive one at that. A deductive discipline is simply not interested in history. It seeks to be mathematically sound, believing its theory to be applicable to any time in history, and it is, in theory. Thus, the eighteenth-century economist philosophers such as Hegel and Marx were no longer considered relevant and basically jettisoned from the field.

Similarly, much of Adam Smith's philosophy was discarded; only certain aspects were retained, such as the oft-mentioned "invisible hand of the market" and his theory of "division of labor."[6] Even Marshall's fascination with history, which undoubtedly motivated him to help found the field, was no longer considered relevant. Today's economists pay little attention to the historical position of the field or of human history in general.[7] But that is not all they leave out. We are just beginning.

Economists also had to employ some capricious assumptions about what was important to them or not, dividing these events into "economic" or "noneconomic" activity. What is deemed noneconomic activity is also referred to as an "externality," much like the "hidden variables" of physics equations.[8] This means that the entire natural world—the air, the earth, the water, and the light—everything that gives us life, is mostly an externality to the field of economics. But that is still not the half of what economics abstracts.

To protect its turf from intrusion from other disciplines, psychology in particular, economists have made sweeping assumptions about human nature. These have been made without much basis, except that it furthers the discipline. Economists are guilty of promulgating the idea that we are insatiably acquisitive consumers, seeking to provide for our individual needs only, not in any way concerned with our family, our relative position in the community, or the larger society.[9] In the eyes of most economists, we are separate individuals who are hell-bent on acquisition, not social creatures infused with caring and compassion. The society, when considered at all, is treated as simply an aggregate of individual desires.

Numerous psychological experiments do not confirm this viewpoint. When given a choice, human beings act with compassion and concern for others a good bit of the time. We care about others, and certainly about how we appear to others.[10] We are very much social animals. In effect, economics remade man in its image and, some say, into its own peculiar species: the species "homo economicus."[11]

But that's not all.

Economists have also made convenient assumptions about the human relationship to nature, reducing all of nature to "land" and "real estate," an assumption that may have helped foster the greed they assumed to exist *a priori*. As Buffy Saint Marie sang, "They put Mother Nature on a luncheon plate, carve her up and call it real estate." This reduction is possible because economics only concerns itself with human actions; everything that does not involve human beings is considered noneconomic, and therefore irrelevant.

Consider what that means for a piece of land. It means that the complex web of ecological relationships that exist amongst all the species that inhabit that land are not salient to economics unless or until there is human involvement. And then, it only concerns itself with a specific kind of human activity, such as the purchase, sale, or development of the land. Again, the quality of the land—and the quality of the water, soil, and air that support the existence of all the life forms living there—are all unimportant; they are mere externalities to economics.

The Economics of Water: A Watery Interlude

Water is arguably our most precious resource, for it is something everybody needs. For millennia, water was thought of as a "commons"; it was something to be shared freely, not the property of any person or species. But water is now in the process of being privatized.

The more common bottled water becomes, the less likely water will be considered a commons. Already, our dwindling supplies of clean fresh water are decreasingly serving the local communities from which they spring, and increasingly being redirected to businesses that bottle it up and sell it elsewhere as a commodity. This business opportunity would not exist if we had not despoiled the water in the first place, of course, and I am ashamed to be part of the generation that let this happen on our watch.

As a teenager, I traveled, by auto and on foot, in the High Pyrenees of Spain. I remember eagerly stopping at various places along the way to drink *agua fresca*. I vividly remember how amazingly pure the water tasted, and I wonder today how many of those springs are still fresh. Unless something changes, water will become the new oil. It already exceeds oil in price per gallon, and by a lot. Many people, including environmental activist Vandana Shiva, predict that one day it will be as contentious a resource as oil. Wars have already been fought over water.

It makes me profoundly sad to see water bottled. The nature of water is to run, and economists of all people should understand this because the nature of the economy is the same; that is why the movement of money is called "cash flow." The privatization of water is symptomatic of an imbalanced and bottled up marketplace, which is exactly what happened in late 2008, causing an international economic crisis. Why do we hold the invisible hand of the market as sacred and allow water to be bottled and sold? Water, like the market, must be allowed to move.

Water is the principle in which life force, movement, is often expressed. Water is life! *Agua es la vida!* This phrase, popular in many languages, needs to be heard, not as metaphor, but as truth. The elements make up all of life, but are not considered alive in themselves. The mind that thinks water is not alive in itself will allow water to be despoiled; the mind that understands water as alive will not.

It is my firm belief that many of the problems of modernity would disappear if we only believed that the elements are alive in themselves, not mere constituents of life. Why? Because if we believe the elements are alive, we tend to respect them more—and we are more apt to build a meaningful relationship with them, as many Indigenous cultures have for millennia, and as Western cultures once did.

What are the ramifications of thinking of water as alive, or of the air as aware? What if all the elements, and for that matter, everything else, is aware and intelligent? Consciousness, intelligence, awareness, and life can be conceived of in this way—as woven into the fabric of existence.

I predict that, one day, and probably within my lifetime, scientists will discover that everything in the cosmos is alive, just like they are today discovering that everything is interconnected. At present, scientists are continually confounded to discover life in places previously considered impossible, such as deep under the cold, dark of Antarctic ice or in the boiling hot springs below Yellowstone Park.[12] Life on Earth thrives in places as hot as Venus or as cold as Mars, yet we continue to waste time speculating about whether life exists on other planets. This is because we focus on the wrong things. Our perspectival consciousness divides the world into categories of living and dead, animate and inanimate, rather than seeing interpenetrating aspects of the same essence in varying states of organization.

Research has shown that what we have deemed "nonliving" acts as if it is part of a living, interconnected whole. Bohm discovered this early in his career when examining plasma, and conceded "it was as though the electrons were alive."[13] The Nobel Prize-winning chemist Prigogine discovered that "inanimate" chemical solutions of molecules appear to communicate with one another, acting in synchronization to form coherent patterns. Prigogine confirmed these results even with random mixes of molecules; the molecules immediately began to self-organize. The logical conclusion is that nature has an inherent capacity—we might even say drive—to self-organize.[14]

The capacity of "nonliving" systems to do things that we attribute only to living systems need not surprise us if only we are able to bring ourselves to a different conclusion—that *all of existence is alive* and aware and interconnected with us—just as Indigenous peoples have been saying all along.

Could it be that we were too quick to dismiss these beliefs as "primitive animism"? I would say so, and I am heartened by new forms of science now developing that essentially reprise animism under various new names: the Gaia hypothesis, organismic philosophy, systems and holistic approaches, to name a few.[15] At minimum, I just do not see the benefit of continuing to divide life from death in the way we have up until now. It would be far more original to bring these aspects together in a way that collapses that dichotomy—that recognizes that what we call life and death, new and old, are interpenetrating. All of existence is alive (or at least capable of becoming what we think of as life), deeply interrelated, self-aware, and self-organizing.

In fact, many things we call nonliving seem in many ways to be more self-aware than humans, or at least more responsive to their environment. They react to changes in their environment instantly and very effectively. Let me return to water to illustrate what I mean.

Leonardo da Vinci carefully examined the qualities of water and filled voluminous notebooks with observations. He found, among other things, that:

> Water percolates through all porous bodies. Against its fury no human defense avails, or if it should avail it is not for long. In its rapid course it often serves as a support to things heavier than itself. It can lift itself up by movement or bound as far as it sinks down. It submerges with itself in headlong course things lighter than itself. The mastery of its course is sometimes on the surface, sometimes in the centre, sometimes at the bottom. One portion rises over the transverse course of another, and but for this the surfaces of the running waters would be without undulations. Every small obstacle whether on its bank or in its bed will be the cause of the falling away of the bank or bed opposite to it. When the water is low it does more damage to the bank in its course than it does when it flows in full stream. Its parts do not weigh upon the parts placed beneath them. No river will ever keep its course in the same direction between its banks. Its upper parts do not impart weight to the lower.[16]

When water does all these things in a seemingly effortless way, is it merely a mechanical reaction to circumstances, or is it a form of intelligence that thinks and makes fluid decisions that embody an active balancing process? I would opt for the latter. If this seems hard to accept, I ask again: Why is water considered alive when it is inside the human body—where it is renamed as "blood"—but outside the human body, it is considered inert or dead? It would be absurd to say that water within us is not alive because we are made up of 70 percent water. But what exactly is the difference between the water inside or outside the body? If none, why is only the water within us considered to be alive? The same could be said for air or any of the other elements. Why do we separate our inner body and outer world in this manner?

All the elements act in similar ways whether they are inside or outside the human body, which is, in part, why human beings are considered the microcosm of the macrocosm. This is something Leonardo da Vinci also realized. In his notebooks, he wrote:

> Man has been called by the ancients a lessor world, and indeed the term is rightly applied, seeing that if man is compounded of earth, water, air and fire, this body of the earth is the same; and as man has within himself bones as a stay and framework for the flesh, so the world has the rocks which are the supports of the earth; as man has within him a pool of blood wherein the lungs as he breathes expand and contract, so the body of the earth has its ocean, which also rises and falls every six hours with the breathing of the world; as from the said pool of blood proceed the veins which spread their branches through the human body, in just the same manner the ocean fills the body of the earth with an infinite number of veins of water.[17]

I love pouring through the notebooks of da Vinci, particularly when he speaks about water. We can learn a great deal from da Vinci about water, and from watching water ourselves. Specifically, we can learn not to come to a standstill in our minds, but instead to flow around impediments and embrace change with grace and fluidity. We can listen, learn, and course-correct much like a river does.

At the risk of overemphasizing the point, water is life, water is us, and until we realize this, we will never change our ways, the ways that are threatening to destroy us now as well. For, if we do not realize that water is life, and continue to imagine that the water outside our body is somehow of lesser importance than the water inside our bodies, we will continue to disrespect the bodies of water on Earth, and we will continue to pollute the water, which only ends up harming ourselves because we are the water that we are polluting. We simply must wake up to this reality in all aspects of life and act accordingly.

Water is our closest relative; we can no longer pretend to distance ourselves from her or to degrade her into a mere utilitarian resource—an abstract other. Water has the right to be; she has the right to exist in the way nature intended, serving all of life. When we extract her energy for hydroelectric power or pollute her with our industrial wastes, we are pursuing our own selfish ends, our short-term cash flow, at the expense of damming up the life force on Earth for the longer term. We do not own the water, the land, the air, the fire. These are simply not ours to own because they are the essence of all life, not just ours.

Chapter 14

Toward an Original Economics

We are still talking about economics, by the way. Remember I said I would make it watery? I only hope I am not watering it down. In all seriousness, it is important to link economics with water, because at its roots, economics is about the fluid transmission of energy. All commerce is an exchange of energy. Have you ever gone to a store to purchase something and became involved in such a good conversation with the cashier that you forgot to take your items with you? Then you know what I mean.

But economics, while fluid, should not subsume water as its product. The most ominous action of economics is its appropriation of natural substances, such as water and fossil fuels, and treating them as man-made capital. This is a fallacy because we do not make water or fossil fuels, even if we do refine them or prepare them in some way for our peculiar needs.

The privatization of water or fossil fuels is highly dubious, if not criminal, and for opposite reasons. One should be freely available; the other should perhaps never be dug up.

The natural hydrologic cycle is a fixed, but ever recurrent process, evaporating water from the oceans up into the clouds and returning it to the earth as precipitation of some kind, whereupon it drains into rivers, flows back to the oceans, and then rises again up into the atmosphere. The cycle continues evermore, and there is always the same amount of water.

Fossil fuels are a different animal; in fact, they are made up of mostly plants and extinct animal matter, created through millennia of heat and compression. It takes a very long time to create oil. In a real sense, oil reserves are nature's oldest graveyards, but we trespass upon and rob from them nonetheless. On the one hand, they are more precious than water because they are natural capital that must be preserved at all costs, much like seeds. On the other hand, they are less precious than water because we can live without fossil fuels, just not in the way to which we have become accustomed. And this leads us to the next pernicious assumption of economics.

Growth: Is It Always Good?

Mainstream economics assumes that growth is always good. It is hard to find an economist who does not believe in economic growth, and the more growth the better. Our obsession with growth is related to our obsession with linear thinking and continual "progress" away from our origins. If an economist is not for maximum growth, they are ostracized from the field, as the controversial economist Herman Daly was, at least until recently when circumstances changed. Both Daly (author of *For the Common Good,* with theologian John Cobb) and E. F. Schumacher, the visionary British economist and author of *Small Is Beautiful,* were attacked for daring to point out something that should be common sense: the ecology of the planet is finite, and we need to plan the scale of human use accordingly. We cannot keep growing the human economy indefinitely.

Schumacher and Daly are original thinkers; their thinking expanded far beyond the boundaries of both their times and their discipline. They spoke their truth and ended up taking their share of arrows in the process. For this, they are heroes of mine.

Schumacher's legacy is the Schumacher Institute and one of my favorite magazines, *Resurgence,* which has now merged with *The Ecologist.* Schumacher's book, *Small Is Beautiful,* has obtained legendary status. The book was a short but effective means of promoting Schumacher's mission, which was to revitalize commonsense economic practices that reimagine the scale of human activity in keeping with the finite limits of the planet. The Schumacher Institute and *Resurgence/Ecologist* work in tandem to affect a rethinking of education and social change for our times by bringing together old and new in creative combinations. They exemplify original thinking.

Daly too has greatly impressed me by rethinking economics in a comprehensive and truly radical way, by going to its roots. Through Daly's efforts, and the efforts of those with whom he partnered, the soil of knowledge was retilled, resulting in a brand new resurfacing of a very old field, now called "ecological economics."

Ecological economics is commonly thought of as a subdiscipline (when it is considered at all), but it could potentially become a *metadiscipline* because it takes a transdisciplinary approach to economic phenomena. It honors the roots of economics while subverting many of its current fallacies, resulting in a fresh—and timely—view.

I do not want to be too harsh on economics because, in truth, its story is only a reflection of the times. Economics did not happen in a vacuum. The story of economics is part of the same story we have been chronicling all along—a shift in thinking from the real to the abstract; from the circular to the linear; from story to information; from diversity to "monocultures of the mind." (The latter phrase is taken from the title of a book by the aforementioned Vandana Shiva, one of my favorite original thinkers. Shiva is a physicist turned activist who is seeking to revitalize traditional agricultural practices in her native India and beyond.)

In short, economics is not unique among disciplines. It did what other disciplines did, carving out a niche for itself in its formative years so that it could operate by its own rules. We could do a similar analysis of most any other academic discipline. The reason I chose to focus on economics is that economists wield so much power in the world and, because they do, they can cause great good or even greater devastation.

Original Solutions for Economics and Beyond

If economics is to right its ship, it must go to its roots for solutions, for it is in the roots where we find the sweet, wet nourishment that can connect economics back to its original sacred purpose. Curiously, because professions or fields are indeed supported by their roots, I wonder if the term field was originally a metaphor at all. It makes me think of a Little Bear joke. He deadpans, "if you want to be outstanding in your field, become a farmer."

In all seriousness, a rethinking of economics might well start with the etymology of the word itself. "Eco-nomics," much like the word "original," has been almost completely severed from its roots. The prefix "eco" comes from the Greek *oikos,* which means "home," and *nomos* means "management." The original intent of economics—what used to be called *oikonomia*—was "management of the home." When I was a child, we could still take "home economics" at school; this was but a remnant of its original meaning.

Economics, of course, also shares its prefix with the word "ecology." This is not a coincidence. Ecology today is taken to mean "the study of ecosystems," but was originally the study of homes or *habitats* before the word "ecosystem" came into use. The combined habitats of a particular place make up the modern word ecosystem.

There is an inherent partnership between ecology and economy. But while a partnership can exist, it is the economists who should yield to ecologists, not the other way around. Economics is a subset of ecology because human economics is always dependent upon the economics (as in limits) of nature, something economists have conveniently forgotten. Originally, economics concerned itself with the place of the human in the whole of nature. It can do so again if the public demands it.

Today's mainstream economics is actually more closely related to the ancient word "chrematistics," an obscure and appropriately ghastly sounding term originally coined by Aristotle.[1] Chrematistics as defined by Aristotle is a "particular branch of political economy relating to the manipulation of property and wealth so as to maximize short-term monetary exchange value to the owner."[2] Chrematistics was frowned upon in the ancient world, but today it has largely succeeded in redefining the purpose of economics and, arguably, has fully succeeded in doing so within mainstream circles.

Once we realize that mainstream economics is really chrematistics in disguise, we understand what happened and what is at stake. The mainstream philosophy has sought to absorb classical economic theory—and its predecessor *oikonomia*—in a way that reminds me of how mainstream modern medicine seeks to absorb complementary medicine (or complementary and alternative medicine, sometimes abbreviated as CAM). The process involves renaming CAM as "integrative medicine" so it can become assimilated into the mainstream, but mainstream legitimization comes at a price. It threatens CAM's original integrity.[3]

The practice of economics as chrematistics cannot continue if humankind is to survive, and thankfully the trend is being reversed by people like Schumacher, Daly, Cobb, well-known alternative economist Hazel Henderson, transdisciplinary thinkers Charles Eisenstein (author of *Sacred Economics: Money, Gift and Society in the Age of Transition*), and Riane Eisler, whose latest book *The Real Wealth of Nations*, reintroduced the phrase "caring economics" to our vocabulary, because we had traveled so far away from the roots of oikonomia.

What exactly is oikonomia anyway? It was defined by Aristotle as "the management of the household so as to increase its use value to all members of the household over the long run." Daly and Cobb have effectively expanded and modernized the definition of household to include "land, shared values, resources, biomes, institutions, language, and history" or, in short, "the economics for community."[4] It is succinctly called "humankind's household" in every *Ecological Economics* journal. (The journal editors also briefly define oikonomia as "nature's household.")[5]

We could call oikonomia "sacred economics," as Charles Eisenstein has, or "caring economics," as Riane Eisler did. But for our purposes, I rename oikonomia "ecological economics" because there is already a journal by that name, and it shares the ancient goal of serving the greater good. I should point out, however, that by Greco-Roman times, the meaning of the phrase greater good had already contracted to signify only the good of the human community. But today, the idea of the greater good is again expanding, due to a resurgence of Indigenous teachings that emphasize our radical interdependence with our plant and animal relatives.

I suggest, then, that ecological economics, infused with the original spirit of oikonomia and Native America, would have these objectives:

- It would take the long view, looking at the impact of actions for seven generations hence
- It would consider costs and benefits to the whole human community and all our relations
- It would focus on real values for actual use only in the amount necessary, rather than abstract exchange values that encourage unlimited accumulation.[6]

Caring Economics

Riane Eisler exposed the caring, feminine side of economics—a side that had been overturned and unnoticed—even as it is implied within the original meaning of oikonomia as management of the household. With understated simplicity, Eisler subtitled her latest book *Creating a Caring Economics* because so much of current economic theory is clearly the opposite: uncaring, unfeeling, hard-number calculations divorced from the actual human experience of suffering, kindness, and compassion. When economists talk about unemployment rates or GDP growth, the impact upon actual human relationships—or the impact upon the web of relationships in the environment—is ignored. Eisler is endeavoring to recast economics in light of the relationships that "support and enhance human life"—and she feels this should be the primary focus of economic analysis.[7]

Eisler's view of economics is that it can only be understood in the context of the myriad of other social factors that shape it—the larger collective of psychological and social dynamics of relations in all spheres of life.[8] She is therefore an effective critic of abstraction in academic disciplines and professions.

Eisler's critique of economic theory is an extension of her lifework, a critique of society and its persistent devaluing of the feminine which, in effect, ignores half of the wholeness of life. Her most celebrated work is the *Chalice and the Blade,* which took a nonlinear view of history and forecast a resurgent feminine ethos. She was on the leading edge of seeing a societal transition from what she termed a masculine-based, "dominator society" to a more feminine-based, "partnership society." This transition is not merely about a shift of power from men to women, but a rebalancing of the masculine and feminine principles. This will occur both within and without, as all of us, male and female, have both a feminine and masculine side.

Economics is no different. It, too, has a feminine side that must come out of the shadows and take its rightful place. Only its masculine face has been acknowledged for much of modernity—at least until Eisler and other visionary economists such as Hazel Henderson (*Creating Alternative Futures: The End of Economics*); Lynne Twist (*The Soul of Money*); Winona La Duke (Ojibwe economist and activist); Julie Nelson (professor of economics and feminism); and others came along.

What Has Been the Role of the Feminine Throughout History?

The fact that the feminine side of economics has been ignored is not surprising in light of the largely unacknowledged presence of the feminine principle in society as well as the marginalized role of the feminine throughout history, particularly in the formative years of humanity.

Eisler followed the groundbreaking example of Marija Gimbutas, the archeologist who asserted that ancient peoples were gynocentric and largely peaceful, enamored with the life-giving, life-nurturing properties of the feminine. Gimbutas formulated her theories based on archaeological research that demonstrated the predominance of feminine deities in Neolithic societies. The archaeological evidence is not in dispute but, not surprisingly, her conclusions came under attack because they threatened the existing paradigm that supposed masculine dominance to have always been the case. The venerable Joseph Campbell, among other luminaries, rose to her defense, saying her research "changed everything," and openly regretted that it was not known when he published his seemingly comprehensive, four-volume work on mythology: *Masks of God*.[9]

Eisler, emboldened by the research of Gimbutas, predicted that the feminine would rise again, and pointed to numerous examples to indicate this renaissance is already underway, if in nascent stages. Eisler was encouraged, not just by the past successes of partnership societies, but also by her own assessment that the current dominator paradigm was beginning to crumble under its own weight.

Eisler was prescient in seeing the end of an era that had basically outlived its usefulness. Consider marriage. As an institution, it was largely established as a way of controlling women through economic oppression, for women were treated as chattel. While this form of oppression still occurs in certain parts of the world and, arguably, to some degree wherever there is unequal pay for equal work, the definition of marriage is rapidly changing. Since the 1980s when *The Chalice and the Blade* was first published, the way we view sex and marriage roles has dramatically shifted in favor of greater freedom and equality, just as Eisler predicted. Consider public opinion on gay marriage. A short decade ago most of society was against it, but in 2014, the majority of (U.S.) citizens and, most significantly, 80 percent of youth, support it, effectively sealing its inevitable acceptance both legally and morally.

This is not to say that the masculine path has ended, but it is time for it to reinvent itself. For too long, that path has been associated with the development of rational mind and the concomitant separation of humanity from nature, which has led to a relentless forging ahead of progressive innovation that has been both remarkable in its accomplishments, but also destructive in too many ways.

The same mentality that helped land a man on the moon, crack the genetic code, and find a cure for polio, has also helped unleash the most devastating weaponry known to humankind. It seemed that weaponry had culminated with the invention of the atomic bomb, but we have since created bombs thousands of times stronger, not to mention an increasing amount of chemical and biological weapons. Cyber-attacks and drone strikes target suspected terrorists while routinely killing scores of innocent civilians, but these are also acts of terror that only perpetuate the cycle of violence. The latter incidents are dismissed as "collateral damage," a term as soulless as the drones themselves. As Macy noted, a large amount of modern innovation is in service to a "military-industrial complex" that is ultimately dehumanizing.

From Domination to Partnership

Eisler focused on the need to shift away from domination and move toward a renewed feminine spirit of life-generating, life-nurturing cooperation between each other and with the natural world. This caring partnership philosophy ought to be at the heart of economics because, after all, oikonomia originally meant management of the home, and one cannot properly manage the home without establishing and nurturing positive relationships. The caring work most associated with women, from child rearing to domestic work, to the professions of social work, nursing, and the like, is either entirely unaccounted for in current economic models, or paid at a level incommensurate with its worth. The idea that we all need to be loved and cared for is not controversial. Why, then, is caring not economically valued by society?

The answer lies within the story we have chronicling all along. The movement away from origin has also been a movement away from the feminine principle, away from the feminine aspect of Earth—her soft, sensuous, mothering support, and toward an abstract, clinical control of her, or rather, "its" resources.

I do not wish to oversimplify the situation because, of course, abstraction has its benefits. There is a benefit to separating nature into parts in order to study a particular aspect more precisely, but this must always be understood as a limited application. It is never the whole truth. We must counterbalance abstraction and separation with an inward movement toward original wholeness and reconnection with all who share our planet.

Rational thinking and intuitional thinking, left and right brain, are two different energies that require balancing. Balance can come about *because of their difference,* without one trying to be the other. A good analogy is an emulsion of oil and water. What is required is a kind of "spiritual mayonnaise" in which masculine and feminine, rational and intuitive, harmonize while still maintaining their separate identities.

At present, we have created a peculiar kind of world through our thoughts—one that is imbalanced toward the masculine forces associated with initiation, exploration, and conquest of new frontiers (both mental and physical places). This initiatory outward energy has an important role to play, but it can easily become ungrounded and headstrong. If so, its energy is spent. How can it then renew? Its renewal is dependent upon acceptance of a counter force of feminine original wholeness, which is more of an inward force that draws one back toward the center, the place of inner wisdom and emotional intelligence.

We are now on the precipice of change. The inward movement toward wholeness is slowly regaining its footing—and not a moment too soon. If we are to reach our full potential as human beings, all of us, men and women, are going to need to harness the sacred feminine energies that have been too often neglected in modernity. This is not only about rebalancing relationships. The relationships between men and women are but a symptom of something far greater and more important—the complementarity of two energies that exist in all of nature: the outward yang and the inward yin, the sacred masculine and the sacred feminine.

Every part of nature has a masculine and a feminine side. This is something I learned from Navajo cosmologists David Begay and Nancy Maryboy, who emphasized this in multiple ways: in the act of writing their doctoral dissertation together (where they wrote and talked through every word of the text); within the content of their dissertation; and during the annual dialogue sessions. The idea that every part of nature has a feminine and masculine side was immediately intriguing to me. And when I looked at life that way, I found it to be true. Moreover, I learned from Begay and Maryboy to look for ever-increasing levels of recursive complexity in nature, and that each of these levels maintain both a feminine and masculine aspect in a dynamic tension of balancing—a dance that is never ending, even as it is complete and whole.

The Partnership of Head and Heart

Thought itself has a feminine and masculine face that is recursively found in various vehicles—not just in the head and the heart, but in the right and left brain, and in the right and left ventricles of the heart—with each aspect having a separate but interdependent function.

The wonderful research of the Heart Math Institute has demonstrated that not only does the heart think, but there is a dynamic, ongoing two-way communication between the heart and brain. The heart communicates to the brain in at least four demonstrable ways: neurologically (through the nervous system); biochemically (through hormones); biophysically (through pulse waves); and energetically (through electromagnetic field interactions).[10] The functioning of the heart has been clearly demonstrated to affect the functioning of the brain, and vice versa. The partnership between the heart and brain is key to our overall health and vitality.

Until recently, most cultures thought of the heart as the seat of thought and consciousness as much or more than the brain—and this is still the case in Chinese and Ayurvedic medicine (from India). In Ayurveda, the heart is considered the seat of the *ojas,* or the finest, most subtle level of the physical body, associated with your aura, your vitality, and your emotional intelligence. If a Chinese acupuncturist tells you that you are thinking too much, she/he is probably concerned about stress to your heart. The remedy: slow down; think loving thoughts; be compassionate toward others; count your blessings.

In short, thought is one essence with two aspects: the feminine and masculine. If we focus too much attention on any one aspect, we become disconnected from the whole. If we consider the two together as equal and complementary energies, we are at one with the dynamic wholeness inherent in all of life. This connection to wholeness is a great blessing, something to be eternally grateful for. We have already spoken about the origin of thinking as thanking. The words came into English from the proto-Germanic *pankaz,* for "gratitude" and "thought," which is related to the proto-Indo-European *tong,* meaning both "think and feel." When we think at the most primal level, what is it that we are feeling? We are feeling the pulse of life.

The Partnership between Thought and Nature

When we think, we are participating in life—at least if we think with feeling and awareness of who we are and how we are connected to all there is. Ultimately, we think not only with our brain and heart, but with every neuron and cell in our entire body, which is in turn connected with the larger body of nature. Our bodies are not really separate, but enmeshed in a larger neural net—an energy system that pervades not just our species, but the entire planet. As the Mayan seer Ac Tah puts it, our bodies are like individual light bulbs that are powered not only locally, but from a larger grid. We receive our energy in the same way we receive our thoughts—from without. There is energy pulsating through us all the time; everything we think or do is ultimately connected to every other being on the planet and beyond.

When we think, we are radically interconnected with all of life. Our thoughts, our memories, our wisdom, all come from the wisdom of nature. This is why all attempts by scientists to locate memory traces in the brain have proven futile. When Grandfather Leon suffered a serious stroke a couple of years before he died, he recovered his memories in the most extraordinary way: he went out to sacred sites and had the land retell the stories to him—because that is how he gained his knowledge and wisdom in the first place. But something else happened to him when he first had the stroke, which we could call a "stroke of insight" (with apologies to Jill Bolte Taylor, who experienced something akin to this and wrote a fascinating book with that title).[11] Grandfather Leon, upon having his stroke, went to the other side and conducted extensive conversations with spirit beings from that realm. And when he returned to the realm of the living, he was able to live the rest of life with renewed purpose and dedication. He later spoke of his insights into death and dying—describing how the life force reverses its spiral of energy at the time of death, among other things—in a radio interview on "Native America Calling" with host Harlan McKosato just days before his own passing.[12] This does not mean that Grandfather Leon did not suffer much of the same frustrations that other recovering stroke victims do. He most assuredly did. But, as was often the case, he garnered great lessons from his life experience and was able to pass them on to others.

One lesson I have learned is that there is a relationship between our thoughts and our health, between our health and that of our community, and between our own community and that of the natural world. We can recover our health through interaction with places that are healthy, because we are then healed by the vibrancy of that natural community. Yet I have also come to realize that many sensitive people suffer debilitating health conditions because they are unable to reconcile or separate their own bodies from the disease within their communities and within nature. I believe this is especially prevalent amongst Indigenous peoples, who originally enjoyed excellent health until their world was upended through colonization, which brought pollution on many levels, mentally and physically. But this instinct to feel pain when nature is out of balance is not wrong—and it is something all of us must learn from. We must take care of the human economy only in the context of taking care of nature's household first. We cannot abuse the Earth for purposes of unmitigated economic growth without hurting all of nature and ourselves in the long run. We are all interconnected with each other and the health of our planet, even if we do not want to face this.

Without enough trees, our very breath is compromised, and we have a hotter and more arid planet. Without adequate fresh, unpolluted water, we will die along with the fish, and many of us are facing this condition right now. Without moist and healthy soil, full of countless microorganisms, we cannot grow healthy plants and vegetables with their full complement of vitamins and minerals. Without the pollination of the bees, now disappearing, an enormous amount of plants and vegetables may also disappear. It may sound simplistic, but it is important to always be thankful for the trees, the water, the soil, the bees, the plants and vegetables, for without them, we could not exist.

Some of us are so wrapped up in our personal concerns that we care about them above all else—even in the face of looming ecological disaster. Sadly, there are people who worry more about their receding stock portfolio or their receding hairline than the receding rainforests. But, thankfully, there are many people who do care, and with every generation, that number increases. The younger generations not only care, they expect you and everyone else to care. They expect change; and in fact, are demanding it. These future generations are like the green buds and sprouts in Abram's story of the King's map. They are starting to break through the artificial veneer they inherited from older generations—the cold abstraction of modernity—to welcome back a season of renewed caring and appreciation for the sacredness of life.

Writers like Charles Eisenstein are a breath of fresh air. His view of "sacred economics" reimagines a *gift* economy; he urges us to remember the perennial spirit of blessing, and not to succumb to the scarcity consciousness that pervades modernity. He asks us not to eliminate material culture, but to rethink our relationship to objects such that we care for them, repair them, make them last.[13] He sounds almost Ghandi-esque in his call to exchange profane greed for sacred caring. There is enough for all of us if we live from blessing, and not from lack. As Ghandi said, "there is plenty for every man's need, but not for every man's greed." It is all a question of scale.

The Importance of Scale

An important aspect of ecological, caring, and sacred economics is proper scale. The Earth may be whole and complete, but it has finite limits. Human enterprises ought to temper their ambitions in keeping with the needs and capacities of the rest of the biosphere.

If we are to provide for future generations, we obviously cannot usurp all of nature's resources for our generation. But that is exactly what we have been doing in the name of growth. Let's face it: the reason why economists and politicians (of any party) like growth so much is that it papers over other failings. When the economy is growing at a good clip, we can lower taxes and/or avoid cutting social programs and still not run up as big a deficit. When the economy is growing, employment is full, and when people have jobs, they keep spending, which becomes the engine of still further growth.

There is only one problem with that kind of shortsighted thinking. A job is not going to mean much if we do not have clean water to drink or clean air to breathe; a job is not going to mean much if there are no more rainforests and the earth is turned into desert; and a job is not going to mean much if the sea level rises and wipes out your hometown. The economy of China has grown at about a 10-percent clip for over thirty-five years. What has been the result? The ecology is in ruins, and by that I mean the air is unbreathable; dead pigs are being found in the rivers; and thousands of rivers have dried up—that is *thousands,* not hundreds. It does not take a seer to know that the people will revolt if the government does not change, and soon.

This is why government regulation, considered the enemy of growth, has a necessary place within an intelligent society. The Clean Air Act began in the 1960s in the United States and was revised along with the Clean Water Acts during the Nixon administration in the 1970s, even though Nixon vetoed the Clean Water bill before it was overridden.[14] Since then, many tweaks and shifts in regulation have occurred, but in general, the air and water, if not so much the soil, are improving, at least in the United States. The Obama administration has been thwarted on much of its legislative agenda, but it has succeeded in raising fuel standards and lowering factory emissions by government mandate, which is a decent start. Let's only hope his successor continues the momentum.

One of the most significant government actions ever in this regard was the banning of the pesticide DDT, which occurred in the United States after Rachel Carson's landmark *Silent Spring* ignited the modern ecological movement. (The book was published in 1962, and DDT banned in 1972.) The problem, however, is that DDT was and is still used just south of the border in Mexico,[15] and underground pollution does not respect borders.

A complex decision for the world today is whether or not to establish an international body that regulates the ecology of the planet. Daly, a former member of the World Bank, is for this, and I at least understand his thinking. Global phenomena such as climate change can be controlled only by a coordinated effort throughout the entire world. The nonbinding resolutions passed through international conferences, such as the Kyoto protocol, do not seem to be working, or are not working fast enough.

I can see the other side also, and have empathy for those that fear global government and large-scale consolidation of power. In the end, I pray that human beings will change not because they are forced to, but because they want to, because they understand the need. For that to happen, consciousness has to shift. It definitely is shifting, but is it shifting fast enough? Probably not. We may very well need some stopgap global governance board for ecological issues, something similar to what we have for economic issues, like the World Trade Organization (WTO). It would not be enough to set up an ecological oversight board; that organization must then seek dialogue with its economic counterparts. Ideally, the governing ecological board would have as much or more power than the economic board. But, at minimum, the two boards need to dialogue in an intelligent manner.

If we remember the roots of economy, we can break the logjam between the supposed opposites of ecology *or* economy, the spotted owls *or* the loggers. The proposed Keystone pipeline (from Canada to the United States gulf region) was portrayed as a choice between jobs or the ecology. This is a false dichotomy. Pipeline jobs are high paying jobs, as are jobs in many environmentally destructive industries, probably because they are so dangerous. But if the society made a choice to value green jobs, or jobs that clean up the consequences of an already despoiled environment, the same or more jobs could be created. This is already happening more than we even realize, as Storm Cunningham (author of *The Restoration Economy*), cultural historian Howard Mansfield, and ecologist Dennis Martinez, among others, have demonstrated.

Cunningham in particular has documented the rapidly increasing restoration sector of the economy. He defines the restoration economy in eight categories: ecosystems, watersheds, fisheries, farms, brown-fields, infrastructure, heritage, and restoration from disaster/war.[16] By this account, nearly 50 percent of the economy is already of a restorative nature, but there is still much work to be done. Cunningham says:

> We are now in the Re century, when redevelopment displaces devel-opment, reconstruction outpaces construction, redesign is more com-mon than design, reuse betters use, repaving keeps more people busy than paving, reengineering dominates engineering, renew dethrones new, and revitalization supplants devitalization. We're finally kicking our self-destructive addiction to sprawling new growth and replacing it with healthy, sustainable regrowth.[17]

A restoration economy is in accord with the overall message of this book. As Mansfield said, "When we bring our loving attention to that which is old, it is not the past we are restoring but the future. Good res-toration saves the future."[18] This quotation, seemingly about economics, is actually about all of life. To save the past in service of the future is to think originally.

I would be particularly pleased if society directed its efforts to restor-ing watersheds and rivers above all else because nothing else could be more important. Only 3 percent of the water on the planet is of the fresh water category and of that, less than three-tenths of 1 percent is free flowing (the rest being trapped in glaciers or polar ice caps)—and a great majority of that free-flowing water is polluted.[19]

The restoration of a river can often be accomplished by simply remov-ing dams. My principal objection to dams is emotionally based, but I make no apology for that. I know in my gut that damming rivers is dam-aging to everything: to the river, to the fish, to the trees and ecosystems around the river, and to human beings too. The nature of water is to run.

When we obstruct the natural movement of water, we harm ourselves because we, too, are made of water. I live in New Mexico, near the Rio Grande, which is a pale facsimile of its former self, largely due to a dam that exists midway between Santa Fe and Albuquerque. The stately cot-tonwood trees that line the river are dying because the river never floods anymore. For centuries, that river provided life, and it is our moral duty to restore her to health, along with all the other rivers we have bottled up. I have seen healthy rivers. When I was in Costa Rica, I was enthralled to feel the life force pulsating from the river and all around the river, where the mangroves grew so lusciously. I could feel the river as if it were my own pulse, my own veins. It was there, while dreaming of my wife's smiling face, that I had this vision:

> I (consciousness) became dolphin. Dolphin dove down deep toward the bottom of the ocean, swimming powerfully and freely. Just before reaching the bottom of the ocean, dolphin realized he was a river dolphin and reversed course, heading to the surface. But when dolphin broke the surface, consciousness continued to rise into the sky and yielded a bird's-eye view of the dolphin as it swam toward the mouth of the adjoining river. Consciousness soared even higher like an eagle, and looked down on many rivers—all leading into one ocean. It became apparent that these rivers were the ventricles of a human heart and, as this awareness became known, a beautiful drumbeat emerged from the heart of Mother Earth. It was the heartbeat of Mother Earth, and everything was in complete and total ecstasy.

When I returned from Costa Rica, I told this vision to Grandfather Leon, and he suggested that the reason why there is so much cardiovascular disease on this planet today is because of the damming of the rivers. "The rivers are Mother Earth's veins," he told me. When they are obstructed, we have cardiovascular trouble, and when they are unobstructed, our human spirit is set free. I am grateful for his counsel, which I recognized as true. There is a relationship between the free and natural flow of water outside our bodies and the health of the circulatory system inside our bodies. As Thomas Berry and others have said, "We cannot get well on a sick planet." Of course, we must restore the rivers to their natural course. Our bodies, as da Vinci demonstrated, are a microcosm of the macrocosm.

In the same way, our thoughts are always an inlet to the whole. The sooner we realize this and think along with the flow of nature rather than trying to outsmart or oppose nature, the sooner we move toward inner peace and harmony. Our thoughts want to flow just like a river wants to flow—effortlessly.

PART FOUR

Renewal: Can Education Promote the Renewal of Original Thinking?

The past is never dead. It's not even past.

—William Faulkner

Why do schools teach almost nothing of the pattern which connects?

—Gregory Bateson

No one is really working for peace unless he is working primarily for the restoration of wisdom.

—E. F. Schumacher

The greatest enemy of wisdom is not ignorance, but the illusion of knowledge.

—Yuval Levin

Imagine a world without hypothetical situations.

—Dan Moonhawk Alford

To learn, One accumulates day by day.
To study Tao, One reduces day by day.
Through reduction and further reduction

One reaches non-action,
And everything is acted upon.
Therefore, One often wins over the world
Through non-action.
    —Lao Tzu

Water Jar Boy

A young girl lived with her father and mother in a very old village near a place where the waters come together. The girl was very beautiful, kind, and good of heart. When she became of age many boys of the village tried to win her eye. But, she was very shy and did not pay attention to any of the boys of the village. She did not want to leave her father and mother since they were quite old and needed her help.

The girl liked to help her mother make pottery, especially the water jars, which her mother made so well. One day her mother asked her to help mix the pottery clay. She went to a spring near the village to get a special colored clay to decorate her pottery. While she was mixing the clay with her feet near the spring, she began to feel very strange in her stomach. The more she mixed the clay, the more clay covered her feet, and the stranger she felt. She stopped mixing the clay and went home. She told her mother how she felt mixing the clay, but her mother thought nothing of it and told her not to worry.

A few days passed, and she started to feel something moving in her belly. She did not want to tell her mother and father. But soon she became very ill, and when her mother felt her belly she knew that her daughter was with child. When the child was born she saw that it was not like any other child. It was a beautiful little water jar. Her father came in, and upon seeing the beautiful little jar said, "It is a special gift, and although we do not know how this happened we must accept it." The girl's father became very fond of the little water jar: and when it began to move and grow he became happy. The water jar grew very fast, and in a few days it was able to talk and roll itself around following him inside the house. One day, the little water jar asked the grandfather to take it outside so it could play with the other children. The grandfather was surprised at the little water jar's request, but he took it out, and soon the little jar was rolling around to the delight of the children in the village. The children became very fond of the little water jar and would wait each day for the grandfather to bring it out to play. The children named the little jar, Water Jar Boy.

One day the young men of the village were gathering to go rabbit hunt-ing. Water Jar Boy announced, "Grandfather, I want to go hunting with the rest of the boys; please take me to where the rabbits are so I can hunt too." The grandfather was shocked at the request and told Water Jar Boy, "How can you hunt, you have no arms or legs; besides hunting is for real boys!" Water Jar Boy replied, "But Grandfather, I am a real boy!" Grandfather decided to take Water Jar Boy to where the rabbits were, and as they were leaving his mother began to cry fearing that Water Jar Boy would be hurt. Water Jar Boy told his mother not to worry; he would return with many rabbits!

Grandfather placed Water Jar Boy near a mesa where he knew many rabbits lived and told Water Jar Boy that he would return to pick him up before sunset. Grandfather then joined the other older men as they set out to gather the rabbits together. Water Jar Boy began to roll around as he saw rabbits pass by. As he was rolling he hit a large stone and broke. Out of the broken jar jumped a very handsome boy. The boy picked up a stick and ran after the rabbits killing some of them and letting others go. As the sun began to set he walked toward the mesa carrying many rabbits on his back to meet his grandfather. As he approached, his grandfather did not recognize him. His grandfather asked, "Have you seen a water jar rolling around?" Water Jar Boy laughed and said, "Grandfather, it is me Water Jar Boy, your grandson!" Grandfather looked in disbelief as Water Jar Boy told how, when rolling around he hit a stone, broke his clay skin, and came out of the jar. "I told you I was a real boy!"

When they came home, Grandfather announced to his daughter and the grandmother, "This is my grandson, this is Water Jar Boy!" Then they told the story of how Water Jar Boy had jumped out of the broken water jar and how he had killed many rabbits. Everyone was happy, and they invited all their relatives for a special feast to meet the new Water Jar Boy. From then on, Water Jar Boy stayed with the young men and participated in the life of the community.

As time went on, Water Jar Boy became curious about who his father was. One day he asked his mother, "Who is my father, where does he live?" She began crying and said that she did not know, and she could not tell him where he could go to find him. But, somehow, Water Jar Boy knew the answer to his own question. He announced to his mother, "I know where my father is, and tomorrow I will go and find him!" The next day he set out toward the west and walked for a long time. He saw a marsh and knew that there was a spring there. As he neared the spring he saw a man dressed in buckskin sitting on a stone. The man asked, "Where are you going?" Water Jar Boy replied, "I am going inside that spring there to find my father." "Who is your father?" asked the man. Water Jar Boy paused and looked at the man closely and then said, "I think it is you that is my Father." To which the man replied, "Yes, I am your father, and I am happy that you have finally come to see me. I came from the inside of that spring; that is my home." Water Jar Boy's father took him inside the spring. There Water Jar Boy met all his relatives. Water Jar Boy stayed in the spring and lives there to this day.

—A Tanoan Indian story [20]

Chapter 15

Education as Renewal

Can Education Promote the Renewal of Original Thinking?

All of our exploration thus far served to till the soil of our minds—to shake up and break apart our limited assumptions—so that we can learn to think in fresh and original ways, ways that are increasingly whole, inclusive, and interconnected. This was in preparation for addressing what I see as the culminating idea of this book: the renewal of our original human purpose through reeducating our minds to receive thought as a blessing, as a whole and complete transmission from Spirit.

We laid the groundwork in Parts 1 to 3 by tracing the origin of thought as coming from blessing (thinking coming from thanking); as coming from origins (particular places); arising with the patterns of nature; and connecting us to the whole of creation. To complete our reeducation, however, we have to further shift our ordinary conceptions of knowledge, education, and nature. Specifically, we have to reimagine nature and the human being as one; we have to reimagine knowledge as something that arises with the cycles of nature (as we did for thought); and we have to reimagine education not as the accumulation of knowledge, but as a renewal of knowledge. The primary purpose of learning, then, ought not to be to acquire knowledge, but to give back, to enter a reciprocal relationship with the natural world that serves to renew nature and the human community together.

The act of educating is, unfortunately, normally associated with anything but renewal. Too often, as Whitehead and others have observed, education is about the transmission of lifeless, abstract ideas. But in this we speak only of modern education and forget that education existed long before there were schools.

Originally, education was storytelling. It told a story of awakening into the hidden wholeness of reality, a story of learning to become fully human. This was once the most important purpose of education and all of life. There were teachers, mainly the elders, but Nature was understood to be the original teacher. There were tests, too. But these tests were not given merely to assign abstract grades. They were initiations into one's life calling.

When we previously examined the question of what it means to be human, we could not possibly have reached a single, conclusive definition because each of our lives is seeded with a unique purpose—a dharma or calling—which unfolds in the course of our lifetime, whether or not we realize it consciously. At the same time, our individual path is derived from a greater pathway—a grand movement of nature's patterns from which our thoughts and perceptions originate. We are inextricably connected to our origins even as we try to break away from them. Despite our efforts to convince ourselves otherwise, we always remain creatures of this land; our feet remain planted on this earth. It does not matter if the Earth is disguised as a concrete slab or carpeted floor. It is still the Earth.

I wish it were a harmless exercise, an innocent fantasy, to imagine ourselves as separate and transcendent beings somehow hovering above the dark soil of earth in our abstract minds. But the actual effects of our chosen separation are tremendously harmful, both for us and other species. During the past millennium, humanity has been trying to disconnect itself from itself—from our essence and from our origins—and in the process we have wounded our soul more than we know. We might as well be cutting ourselves with a knife when we cut ourselves off from our coevolution with the totality of life on Earth. The whole of creation is the source of our well-being and has been so since time immemorial. It is only through connection with all of life that we can renew ourselves as human beings. Like Water Jar Boy, we are most alive when we realize the depth of our relationship with the Earth and all the elements that spring from her.

I was recently in a beautiful canyon known as Canyon de Chelley (pronounced *de Shay*). Located on Navajo lands in present-day Arizona, the canyon has been comaintained by the Navajo people in partnership with the U.S. Park Service for over sixty years. But such knowledge disappears once you are in the canyon, as does any sense of linear time. I have been to this canyon many times, and I always feel nurtured there. It is as if I am in the womb of creation. It is an environment of water, wind, and eternity.

Entering this canyon, like any canyon, feels like a journey into the heart of time. But I prefer not to analyze it as a geologist might; I simply breathe in the presence of the ages. It is no wonder that sand sifting through an hourglass is a powerful symbol of the passage of time. Inside a sandstone canyon, you feel the organic quality of time. Time is stone. It is stone caressed by wind and water, seduced out of its hardness into sand and soft, wet clay. The whole experience is one of continual birthing. I feel reborn and renewed.

Have you ever noticed how the crevices in a canyon, if followed, inevitably lead to water? On this day, I followed the channels in the rock until I was led to an alcove where the early spring waters had just arrived. I had to crouch low to enter. Once inside, I listened. The gurgling waters spoke first and seemed to welcome me. I gave thanks to the water. Gradually, I expanded my listening outward and upward. The cottonwoods sang a song through their leaves while, a thousand or more feet above, ravens and hawks danced in harmony, riding the upper crest of the same winds. The above and the below, the past and the present, were one. That evening, I had a vivid dream.

> Several of us are in a sacred site—a deep canyon. We are actually underneath the canyon floor emerging as corn through the rock. We emerge through layers of rock and water into the lush canyon floor. I sense time slipping. We are connected to everything in this original source of emergence. We relive the original emergence and at the

same time are aware of present day reality as people. The sandstone itself is time; it allows for passage between the worlds. We emerge through the rock like corn people and are renewed.

The Tapestry of Original Education

I speak of original education because we have forgotten the original meaning of education—its essence—which is hidden in the fabric of life itself. Mother Earth is continually self-learning by weaving a wonderfully inclusive tapestry, and this tapestry contains all the elements and all the animals, plants, and minerals that can be found. The Earth is already educating herself into wholeness. So should we.

The collective actions of nature—reusing waste as a resource; diversifying and cooperating; relying on rich feedback loops; optimizing rather than maximizing—are some of the ways in which nature renews (educates) herself into wholeness.[1] But the wholeness of nature is a dynamic process that is never finished. There is always incompletion in nature, which is why humans can never know everything. In recognition of this, the Navajo traditionally leave at least one imperfection in each rug they weave.

The tapestry of life may be dynamic and unfinished, but it is also tightly woven. There really is no such thing as a completely separate organism in nature. Consider monarch butterflies. They are dependent upon milkweed to lay their eggs; the milkweed is dependent upon insects for pollination; the insects are dependent upon milkweed for food and reproduction, and so forth. Organisms exist in intersecting webs of unfolding relationship.[2]

Human education was once conceived within this larger context of ongoing creativity. The human instinct to learn was seen as a subset of the instinct of all of nature to learn and grow. There is no better or more original teacher than Mother Nature. We still have much to remember from her.

The ancient Greeks, as we have already seen, understood that nature was the primary source of knowledge, and that our task as human beings was to remember—which literally means to become whole again (the opposite of being dismembered) by reuniting with what we were always a part of, but have forgotten. Our memories are tied together with that of Mother Earth. She, in turn, is tied to a greater memory—the entire cosmos. The same elements that exist throughout the cosmos exist on Earth, and in our own bodies. The whole is within us, and we are within the whole.

It was once commonplace to see the cosmos as an integrated and interconnected whole. This understanding has not been lost, but only forgotten, and it can be restored through the revival of perennial wisdom and, in particular, Indigenous wisdom.

The reason we are experiencing a resurgent interest in Indigenous ways of knowing (and of the similarly wholistic roots within Western traditions) is because, on some level, we know we are wounded, and we know we must become whole again. The good news is that the old ways have never completely disappeared, even if they went underground, particularly in the West, for self-preservation. Ancient wisdom from Greek, Druid, Celtic, Egyptian, Indian, and other cultures has continually resurfaced in multiple guises: the romantics, "transcendentalists," theosophists, and more.

This cyclic return of perennial ways of knowing is sometimes obscured because modern movements tend to dress up the old in new clothing. We may not recognize it as a return but, nonetheless, that's what it is. The wholistic education movement, for instance, is said to have begun in the 1970s and appears to have been new at that time. But, not coincidentally, this was also when Native studies programs and tribal colleges began making headway. The two movements are connected. In sum, seemingly new programs are not really new—they are a resurgence of old ways of knowing.

In the following pages, I explore how modern education can renew itself, thereby becoming a true means of renewal for humanity and the planet as a whole. I first look at childhood education, because children are our future. But I also focus on higher education because as we have already seen, higher education affects the professions that affect much of the worldview that pervades modern life.

Chapter 16

Childhood and Education

What Is Education?

The word education comes from the Latin root *educare*, which means to "draw out" or "lead forth," as in bringing forth one's own latent potential. The Sanskrit word *srj* has a similar meaning, implying the organic emergence of creation, as in a spider generating her web from her inner being. These words serve to remind me that a child draws upon their own creativity not in separation from others, but in connection with the rest of nature.

In contrast, the word "teacher" evokes something different because it has come to imply one who imparts information to another who does not know. But there was once far less of a dichotomy between teacher and student; in fact, the word teacher comes originally from the Old English *laeren*, source of the modern "learn" or "lore." These words are clues to a hidden narrative: the evolution of education as it has migrated away from story and mutual learning to the inculcation of knowledge by an authority figure.

Of course, the best teachers realize they are not merely instructors, but guides and facilitators. They recognize that their responsibility is to draw out the brilliance already within the child.

To a child, all of nature is wondrous and awe-inspiring. The essence of childhood is creativity; if allowed to play in nature, all sorts of creative learning will occur. What is truly educational awakens a person to realize their own gifts—what they were born into this world to do. A good teacher is therefore, in many respects, not a teacher at all but a wise elder or muse. She or he is able to evoke something from the student that is already there, perhaps dormant—a seed ready to be watered, nurtured, and blossom.

We in the West typically allow children to be children for a time—that is, to play in the fields, to romp with butterflies, to have imaginary friends. We allow and encourage them to do these things—except when in the classroom. In fairness to many wonderful teachers, I may be somewhat overstating the situation but, invariably, even the best of teachers are constrained by the system they work in. This is true even in schools deemed "progressive" or "alternative."

Why Is Our Current Model of Education Not Working?

The prevailing models of education are outgrowths of the way we have come to see the world. Because we see the world as fragmentary—divided up into distinct things—we have created different subjects to mirror this worldview. And because we divide reality into separate variables that we must control in order to reach reliable data about our experiments, we do the same to children. Children become the variables we manipulate to get our results. The assumption is that children must be taught rather than supported to express their own inner creativity. And we further imagine that children will not learn anything unless we control them.

Of course, many teachers vehemently disagree with the notion that they mainly control children. They may be correct, but it is more likely they do not realize how the system is set up primarily to control children. The underlying reality is that children are perceived as the wild card. The school building is the same year after year, as are most of the textbooks and teachers. But the children change, and so to control change, we control children.

In Waldorf and Montessori schools, developed by Rudolph Steiner and Maria Montessori, respectively, the system is considerably better. There is an awareness of the value of learning directly from nature about the interconnectedness of all things. The students do not use textbooks and are allowed to play freely outside under adult supervision. In Waldorf schools, children are allowed to embrace "imaginal learning," including pictures, stories, and creative play, at least until about age seven or so. But even this is because developmental theorists, such as Piaget, have told us that a child cannot develop conceptual learning—that is, rational thinking processes—until this time. So even in Waldorf and Montessori schools, there is still the idea that children grow up by developing their capacity for rational thought.

The main difference between these and most other schools is that the children of Waldorf and Montessori are not rushed prematurely into an adult worldview. These children are allowed to develop fuller capacities of perception along with the eventual development of rational thinking. Thus, when rational thought is developed, it is (potentially) still rooted in the child's ability to observe living nature.[1]

This is actually a throwback to the origin of rational thinking. "Rational," as noted earlier, is derived from *ratio,* or the observation of beauty and proportion in nature. When rational thinking is seen this way—the original way—it is present and life-affirming because it remains interconnected with emotional, intuitive, and other aspects of thinking.

What Does It Mean to Be an Individual?

One of the key assumptions within Western education is that each child is a separate individual. Although I do not disagree with this notion, I feel it is necessary to take a closer look at what the concept of individuality means, for individuality is not as universal a concept as might be assumed. For instance, one of the key differences between Western education and Indigenous or Eastern education is that the West is intent on developing the ego of the child.

The concept of ego assumes that thinking occurs in separation from others—that we are a separate self-awareness that then projects that individual awareness upon the world. But is this really so?

In actuality, the uniqueness of each child is not created in separation from others. The earliest experiences of childhood are an immersion into a field of feeling—a totality of awareness. All of nature is experienced as aware, alive, and responsive. The child's sense of self develops gradually over time, but not as imagined by Western psychology, at least not at first.

The initial sense of self is not what we think of as an ego identity, but a fluid locus of awareness within an aware world, a sense of being alive in a living universe. This kind of sensuous immersion is hard for us adults to imagine, but, as we have seen, even adults retain this kind of presence in our dreams. That is why our dreams feel so immediate. But when we awaken from dreaming we forget how to perceive directly; we grab instead for a blanket of rationality to cover over, or explain, our direct perception. We do this unconsciously because this is simply how we think; we do not question the distance our rational mind puts between us and our experience of phenomena. We simply believe that is the way it is.

In short, adults need to relearn how to think directly. Children, on the other hand, already know how. A child experiences reality quite differently. Everything in nature is part of a collective living awareness—every blade of grass, every tree, rock, plant, or flower. Over time, the child will develop a sense of distinctiveness—of being in one's own body. But a child's body is not initially experienced as a barrier; it is more of a *membrane,* a sensitive intermediary that connects us to the natural world much more than simply protecting us within our skin. We experience our bodies, especially in early childhood, as highly receptive instruments used for touching, hearing, smelling, and tasting the body of nature. "We are mistaken when we assume that consciousness is an interior human trait that first unfurls inside a young child, who then learns to attribute that same quality to other persons, and … onto the surrounding world of things and beings," writes David Abram. He goes on to explain:

> The self begins as an extension of the breathing flesh of the world, and the things around us, in turn, originate as reverberations echoing the pains and pleasures of our body. So the clustered trees, the bricks on the floor, and the sunlight are not first encountered as inert

or insentient presences into which, later, the child projects her own consciousness. Rather, the inwardly felt sentience of the child is a correlate of the outwardly felt wakefulness of the sky and the steadfast support of the ground, and the willfulness of the caressing wind; it is a concomitant of the animate surroundings. Only much later, as the child is drawn deeply into the whirling vortex of verbal language— that flood of phrases that earlier surrounded her simply as a beckoning play of melodic sounds continuous with the cries of ravens and the rumble of thunder—only then is the contemporary child liable to learn that neither the bird nor the storm are really aware, that the wind is no more willful than the sky is awake, and indeed that human persons alone are the carriers of consciousness in this world. Such a lesson amounts to a denial of much of the child's felt experience, and commonly precipitates a rupture between her speaking self and the rest of her sensitive and sentient body.[2]

The story Abram relates speaks to a particular aspect of education that is rarely recognized. Modern childhood education is not so much about learning as it is about unlearning; specifically, we unlearn a way of being in the world in favor of a way of thinking about the world. We systematically drum out of a child's awareness the kinds of supersensible knowing that all children share. It is hard for us adults to admit, but children possess many faculties that most of us lost long ago—faculties of intuition, spontaneity, and a felt connection with the natural world. Children also commonly display abilities that we adults consider to be extrasensory perception: telepathy, telekenesis, astral travel, lucid dreaming, and perception of energies or spirits in the land—even bilocation, time travel, and past life memories, in certain cases.

Children can do amazing things right up until they learn that they cannot—that is, until they are chastised for suggesting they can. Because adults have lost most of these abilities, it makes us uncomfortable to admit our inferiority. I once attended a quintessential California spoon-bending party. Everyone was loudly chanting "Bend, Bend, Bend," and I found it rather strange at first. But then I saw with my own eyes how easily children could bend spoons with their minds, which inspired me to successfully do it myself—but not like the children, who were able to bend much larger metal rods. The children were able to bend metal with their minds because they believed it was possible. It probably did not hurt that the adults in the room at least wanted to believe, and so were supportive.

Adults do not necessarily suppress childhood brilliance out of jealousy—at least not consciously. We do so because we have a superior sense of reason, and so we teach what we know. But what would happen if, instead of suppressing these extrasensory abilities, we encouraged their development? Children would then become remarkably gifted adults, maintaining a much wider and more generative worldview. They would not be conditioned to narrow the range of possibilities for their lives and for the planet. Buckminster Fuller, lauded as a great visionary, once said he did not possess any faculties beyond those of ordinary children. The only difference was that he actively cultivated those gifts whereas most adults are eager to dispense with them, or have forgotten them long ago.

Can adults recover at least some of these child-like ways of direct seeing? It may be difficult, but I would say: yes, we can; but we have to first believe it is possible and then make a commitment to doing so.

Is There an Ego?

In Western culture, a sense of ego and the ability to reason are developed in tandem. Both teach us that we are separate from nature.

In order to relocate awareness away from a living world and into our heads, into our individual ego, we are taught to muffle nature's voices. This process can be confusing for children. A child experiences the call of a bird or the caress of a breeze as immediate, alive, and aware; the concept of ego, on the other hand, is an abstraction. A child cannot understand it. All a child knows is that adults are trying to get him or her to form some kind of "me-ness"—some kind of identity around the self that is different from everything else. This undoubtedly produces a trauma—a sense of being cut off from a previous sense of nourishment, a sense of being embedded in the arms of nature.

Children are resilient, of course, and they quickly get over this confusion and learn to fit into adult's conceptions of what we want them to be. But I am sure there is residual harm in severing a child's connections with nature. It causes the child to unlearn their integral participation in the natural world, and this takes away a source of renewal and regeneration, which is our birthright as human beings. Our children have become sufferers of "nature deficit disorder" as Richard Louv, author of *Last Child in the Woods,* put it, with far-reaching implications for their mental, emotional, and spiritual well-being.[3] But if we have unlearned our original connection to nature, can we not also unlearn our sense of individual ego? After all, ego is only an illusion, is it not?

It is actually a thorny philosophical question as to whether or not there is such a thing as ego. The ego may have originally been an illusory abstraction, but it has been invoked so many times by now it has been reified into existence. In other words, it exists because we believe it does. But whether ego is real or not, the corollary effects of our belief in ego: fragmentation, anxiety, separation, loss—are undeniably real. Moreover, they are potentially devastating for self, society, and nature. In short, once we believe in an ego, we build a separate identity. And this identity reinforces the illusion that we are all somehow apart from each other and apart from nature. But we do not see this as illusion; we see it as reality because our thinking processes themselves are unconsciously fragmented.

Bohm's caution bears repeating here—that "fragmented thinking creates a fragmented society." We identify with our thoughts, seeing them as ours alone, and this makes us feel isolated. We never lose our need to belong, and belonging was once a given in our place of origin. But now that we have become derooted, it has become a vague yearning, a restless "murmur in the night"[4]—a dissatisfaction we never completely understand. And so we go about awkwardly trying to reconstruct a sense of belonging by joining our fragmented identity with other fragmented identities.

The extreme manifestations of this phenomenon are violent gangs, where troubled youth are willing to kill others or risk their own lives for a semblance of belonging. But many, if not all, of society's groupings are gangs of a sort. They join wounded individuals with other wounded individuals—fragments to fragments. Even nation-states are manifestations of this modern phenomenon on a larger scale, for we can never recreate a coherent whole after first tearing it to shreds, and nation-states were only created after we systematically destroyed many of our tribal roots.

Contrast our modern society with ancient societies, where the primary identity was tribal, not individual. There was no belief in ego within these societies, and thus, arguably, no ego at all. Indigenous societies still employing their ancient languages are a window into these old ways of thinking and knowing. We saw earlier how contemporary Blackfoot language is a language devoid of even the concept of "solids within solids"—let alone the concept of a separate ego.

In many traditional societies, a sense of identity is constellated around a fluid and ephemeral center, something like the eye of a hurricane or the center of a whirlpool. The center of the vortex is an ever-changing "no-thingness" that relates to the flux of all manifestation—the things of life. In Mahayana Buddhism, this is known as *sunyata*. Out of emptiness comes fullness; through nonegoic attachment to things, there is open and spontaneous relationship with all things. This is why Little Bear reminds us in virtually every dialogue that we must empty our cup to make room for new thoughts. This is the essence of open and original thinking. It also captures the essence of what I mean by "renewal through education": renewing our relationships with the rest of creation.

In reality, we humans are no less interconnected with nature than the milkweed plant. We depend upon the same elements to live: light, air, water, and earth. In fact, we are far more interconnected with the rest of life than we typically imagine. Our DNA is our unique genetic signature, but current research suggests that our DNA is between 98 and 99 percent the same as a chimpanzee, and about 50 percent the same as a banana.[5] We, particularly in the West, however, persist in imagining ourselves as separate beings, and by separate I mean separate bodies and separate everything, but, in particular, separate thoughts.

It is not only Indigenous societies that understand thought differently. In Japan, where I lived for close to four years, the concept of an individual psyche in separation from all other psyches is completely foreign. I once visited Tokyo University and was surprised to see that the entire psychology library (at least the books in English) did not include anything other than what we in the West would call "social psychology." For the Japanese, the idea that we exist as a separate individual within a psychic shell of consciousness is nonsensical. The Japanese word for human being is *ningen,* which means a relationship between self and others—not a stand-alone individual. The Japanese sense of self is similar to that of Indigenous peoples: one's relationships matter most. This mindset helps maintain a coherent and cohesive society, and in fact, Japan is still a coherent society, as evidenced most recently in the aftermath of the 2011 earthquakes and tsunami. Of course, the shadow side of cohesion, if carried to an extreme, is insularity and a feeling of superiority. This happened in Japan during World War II.

Western mainstream society, on the other hand, has promoted a culture of independence and freedom and thus created a different shadow, one of separation and alienation. We have become a largely incoherent society in no small part because we have adopted the myth of the individual as separate from society. Alan Watts put it effectively if bluntly:

> When we are children, our other selves, our families, friends, and teachers, do everything possible to confirm us in the illusion of separateness—to help us be genuine fakes, which is precisely what is meant by "being a real person."[6]

The Western worldview, which psychologically separates us from both other humans and from all of nature, is ultimately untenable. To the extent that we succeed in separating ourselves from nature, we rupture our soul.

Chapter 17

Higher Education

How Did the Modern University Become the Way It Is Today?

In ancient Greece, the first schools begun by Plato were academies because they were based in Akademia, a place that originally was a sacred grove of olive trees dedicated to Athena, the goddess of wisdom. The spirit of Athena permeated these early academies, which were wisdom schools designed to foster free inquiry into the whole of nature—the "natural philosophy" of old.

In time, these institutions became known as "universities" (university coming from the Latin *universalis*, related to universe). The emphasis remained on universal thinking, that is, understanding the whole. Today, however, that emphasis on wholeness is lost, due largely to the proliferation of discrete disciplines, each with their own distinct perspectives. Although this trend has almost completely severed us from any sense of wholeness, it is actually a very recent development. In fact, as recently as 200 years ago, there were typically only four disciplines taught in major universities: law, medicine, physics, and theology.

At first, theology anchored education because theologians, who comprised the learned class, founded and staffed the first modern universities. But, over time, universities changed dramatically, abandoning elitism in favor of expanding markets. Universities sought to serve the rising needs of industry for educated workers, particularly in the United States, and so they became increasingly secular. This trend continued in the late nineteenth and early twentieth centuries as university leaders strove to keep pace with the demands of the industrial sector. Over time, the academy increasingly mirrored industry. That path is characterized by two trends that interact and reinforce each other: professionalization and specialization.

The ongoing process of separation changed the nature of work. Whereas once many workers were craftspeople, engaged in the creation of whole things, the industrial age gave rise to specialization, wherein workers only focused on parts; the model is the factory where workers on assembly lines specialized on the separate parts without a view to the whole.

As these processes became embedded in the workplace, they also began to shape the larger society. More and more separations began to occur, taking the form of newly distinguished fields and market sectors as well as the rise of professionals with specialized expertise. The professionalization of the academy itself was not long behind.[1]

Specialization catalyzed the proliferation of a wide variety of new academic disciplines. From the core disciplines of physics, biology, and the human sciences, universities began a rapid, amoeba-like birthing of new fields, subfields, and specialties within fields—again, with the resulting diminishment of wholeness. The change was so rapid and complete that, early on in the process, when Emerson remarked to Thoreau that "all branches of learning were taught at Harvard," Thoreau reportedly replied that yes, "all of the branches, but none of the roots."[2]

How Do Paradigm Shifts in Consciousness Affect Universities?

The short answer is: slowly. It takes a long time before new discoveries are translated into the settled knowledge of the mainstream. It is not just that the ideas being taught are old—which could be a good thing—it is that they are inert and calcified, not alive and growing organically.

In 1925, the famed mathematician and philosopher Alfred North Whitehead made an important observation: that the twentieth-century educational institution was actually based upon seventeenth-century science. This pained Whitehead, because he was acutely aware that twentieth-century physics (both relativity and quantum theory) had changed everything—that new findings pointed to the universe as an undivided whole. Of course, as we have already demonstrated, these new discoveries were essentially rediscoveries of old wisdom, ancient ideas of a radically interconnected cosmos. But rather than dwell on this now, let's focus instead on the justifications for Whitehead's concerns and vision.

Whitehead's fervent desire was for education to be an organic process in keeping with the emerging scientific view of a cosmos of incessant flux and change. His lament was that modern education had become the transmitter of "lifeless abstractions," particularly those which were prevalent in seventeenth-century physics. It was indeed in the seventeenth century when the universe was first posited to be a great machine. (Previously, it was seen as more akin to harmonic spheres of interconnection.)

Whitehead's chief complaint about modern education was what he called "the fallacy of misplaced concreteness."[3] In this view, our partial, abstract perceptions of reality are mistaken for reality itself. As we have chronicled, this way of thinking permeated modern consciousness with the advent of linear perspective. Perspective freezes in place what was previously thought of as animate, as living. Whitehead cogently stated, "perspective is the dead abstraction of mere fact from the living importance of things felt."[4]

There were, of course, benefits to the scientific revolution and the so-called Enlightenment that followed. Our abilities to dissect the world were utilized in the application of impressive technologies, as we have discussed. But the worldview we developed is deeply illusory, because we are never really separate from nature, as Indigenous peoples have long understood and twentieth-century physics was to rediscover. Nonetheless, the illusion of separation paved the way for the Earth, previously thought of as a living mother, to be seen as inanimate and a mere storehouse of raw materials. And this, of course, allowed the Earth and all its resources to be put in service to the rising industrial age.

The age of industrialization, with its emphasis on specialization and materialism—coupled with seventeenth-century science—paved the way to the modern university as we know it today. The history of education mirrors the history of humanity—only it lags far behind, which is why Whitehead was mostly correct in his lament.

What Is the Problem with Specialization?

The argument for specialization is that it allows for concentration on a smaller field of inquiry so that the specialist can advance knowledge in that area undistracted by other fields. There is some truth to this, because an undivided exploration into the wholeness of nature can be overwhelming and so, when taking on a large project, it seems natural to break it down into smaller steps. But unabated specialization without stepping back to see the big picture is beyond foolhardy; it is dangerous. For example, when a person becomes seriously ill in modern society, they are often evaluated by numerous specialists. This in itself would not be dangerous—but the likelihood of little or no communication between the specialists is.

This is most significantly not the way the natural world operates. The leaves of a tree appear to specialize in absorbing sunlight, but they do so far more responsively and interactively than a comparable man-made invention, such as mini-blinds or a solar array. Some tree leaves tilt one way and others swivel another as necessary, optimizing exposure to the sun as it sweeps its arc across the sky. Plant technology operates as a highly efficient solar-collector array, gathering 95 percent of solar energy, a far greater percentage than any human-made array to date.[5]

To understand the natural world, we need to adopt nature's methods as much as possible when we study her. The natural world can and ought to be looked at from many perspectives at once. This is much closer to the big picture of nature. When nature is compartmentalized into separate subjects, however, our thinking becomes fragmented and reduced in scope. The big picture is lost.

Interdisciplinary Studies

Interdisciplinary studies were meant to be the antidote to specialization. The inter-mixing of diverse studies is based in the ancient understanding that the cosmos is ultimately whole and that any single discipline cannot, in and of itself, provide an adequate understanding. However, the modern application of interdisciplinary studies is quite new, emerging shortly on the heels of the establishment of disciplines in their present twentieth-century form.[6]

The movement grew out of a genuine desire for a reconciliation and reconnection of separate disciplines; however, it is a dream that has rarely been fulfilled. For one, funding for genuinely interdisciplinary work is almost nonexistent, according to William Arrowsmith, who worked for years as an evaluator for a major foundation.[7] For another, interdisciplinary work exists in a kind of no-man's land outside the borders of accepted disciplines; it lives in the shadow of academia, not fully accepted or valued.

Most creativity and innovation occurs on the fringes of academic disciplines. This might lead the reader to believe that interdisciplinary work has been a tremendous boon to existing disciplines. But you would be wrong. The borders of disciplines are protected by vigilant sentinels— "knowledge goalies," to coin a phrase—who protect the discipline from infection by new or different ideas. Because new knowledge cannot get in to fertilize existing disciplines, what tends to happen is the creation of even more new, interdisciplinary specializations. The ironic result? Rather than creating more wholeness as intended, interdisciplinary efforts tend to increase fragmentation.

Academic Research

By the twentieth century, the academy had become the place where research is conducted, and this is directly tied to the funding of programs. But the emphasis on research in academic institutions is a surprisingly recent phenomenon. A little over 150 years ago, research and publication were not encouraged at all in academe.[8] In 1857, the Columbia University library was open only two hours a day and, at Princeton, the library was open once a week for an hour! Yet, today, academic work without research seems inconceivable; for faculty, it is "publish or perish." Why the sudden change? One explanation is that the faculty role has become professionalized, mimicking industry, and publishing serves as a measureable way to evaluate faculty. Another explanation is that research solidifies the modern predilection toward accumulating disconnected mountains of knowledge.

The emphasis on research adds to fragmentation because funding tends to go to research that perpetuates the disciplines as separate, professionalized fields. It thus rewards staying out of touch with what is occurring in neighboring fields. It also rewards technical and literary expertise over firsthand experience. A professor was once appreciated solely on the basis of his or her ability to inspire students. Those days, if not gone, are largely forgotten.

Looking back, one can see why seventeenth-century science was—and is—so seductive to commerce, as its principles were (and are) well-suited to the exploitation of the natural world for material gain and industrial invention. And one can also see how it is that the modern university, still eager to serve its alliance with industry (not to mention government-funded military research),[9] has not leapt to fully embrace a different set of assumptions, those of radical interconnection, principles that are both ancient and aligned with modern physics.

Can Modern Higher Education Change?

As I see it, modern higher education no longer has the option of staying the course. It is becoming increasingly obvious that the status quo is no longer working because the economics have shifted. The "professional diploma mill," as Wilshire put it, no longer produces the jobs it used to, even as the cost of education continues to rise. Mark C. Taylor, Chairman of the Religion department at Columbia University, assessed the situation rather bluntly:

Graduate education is the Detroit of higher learning. Most graduate programs in American universities produce a product for which there is no market (candidates for teaching positions that do not exist) and develop skills for which there is diminishing demand (research in subfields within subfields and publication in journals read by no one other than a few like-minded colleagues), all at a rapidly rising cost (sometimes well over $100,000 in student loans).[10]

How did we get in this predicament? We did it by increasingly narrowing the focus of degree programs, by "learning more and more about less and less," as Buckminster Fuller liked to say.[11] Mark C. Taylor's assessment of the closing job market is unquestionably true for the social sciences, even as it may be overly pessimistic for the hard sciences where qualified graduates still find ample openings. But what disturbs me most of all is the decrease in intellectual creativity that has occurred as a result of an overly fragmented educational system.

Unfortunately, the present trend in academe and professional life is still toward greater and greater specialization, which produces a high level of expertise in individual fields, but poor understanding of the whole. This occurs on a routine basis, below the radar, because each specialist is considered responsible for a particular slice of existence independent of everything else. These specialists rarely communicate with their counterparts—other specialists—and instead remain in isolated groups that protect their turf by developing their own jargon, tantamount to creating a new and separate language. This makes it nearly impossible for them to communicate outside their discipline even if they wanted to.

Specialists in a particular field become thought of as experts over time and wield authority over upstarts. There are, of course, the occasional breakthroughs amongst newcomers, such as Einstein. But Einstein, once established, confronted the likes of Bohr, then the upstart. And so it goes. All this behavior is so embedded in our culture that we rarely notice it or realize how recently it all came about. The last point is significant. As discussed shortly, the status quo is not so established as it pretends to be. In fact, the partitioning of the world into the subject areas of academe today—which seems timeless—*is barely a hundred years old.*

Chapter 18

A New (and Ancient) Vision

How Did We Learn in Antiquity?

Ancient ways of knowing did not, of course, divide nature into separate subjects. All of learning was about "seeking life and becoming complete" according to Gregory Cajete, a renowned Indigenous educator. Wholeness is healing. The words "heal" and "health" are related to the Old English *hal* and the Old German *heil,* or "whole." The modern world has run away from wholeness, and as a result, we are unhealthy in spirit. To be in kinship with the whole of nature is healing. It is for this reason that a resurgence of the old ways is important.

In ancient times, we learned by becoming intimately familiar with our environment—that is, we lived with our surrounding environment as family. We did not just take Earth's resources; we made offerings of reciprocation. These offerings, typically corn pollen, tobacco, or something else that we considered sacred, were accompanied by good thoughts. Thought itself was once understood as a form of reciprocation.

Originally, we gained insight through humbleness and openness to spirit rather than through the accumulation of knowledge. We asked for help in navigating the realms that are beyond human sensing in a way that today we would call prayer. Prayer, as I understand it, is original thinking; it is the purest form of thought that coarises in communion with spirit.

It is my prayer that we awaken to our true potential as human beings—that we go beyond the self-imposed limitations of rational mind to embrace the universal Mind that moves throughout nature. On some level, the human soul recognizes the need for this; we know that remaining apart from nature damages our spirit. We long to wrap nature's skin around our own, to submerge our ego and merge with something greater. Rudolph Steiner once observed that what we call spirituality is almost always a desire to dissolve the dichotomy between self and world. He wrote, "We can only find nature outside of us if we first know her within us."[1] This makes sense to me. For until we really feel, deep inside us, that we are the earth, we are the water, we are the air and fire, we may never care enough to change our behavior toward the planet, which is really also ourselves.

Let's face it. Today, humanity is in a critically wounded state. Instead of pretending otherwise, we need to muster the courage to reach deep into our wounds, into the pus and blood of our infected souls. All of us are wounded in some way. We are wounded because the Earth is wounded; when we harm the Earth we harm ourselves. And we have been ruthlessly plundering the Earth for several centuries now. In doing so, we have thrown our biosphere so severely out of balance that some worry we will kill all life on Earth. That is actually not possible; we do not have the power. Mother Earth will survive, with or without us. This does not mean that human beings will survive; we are quite capable of causing our own destruction. We have already taken down many majestic species, or nearly done so, as with the bison and the whale.

In order to recover from the pollution that we have inflicted upon ourselves and the Earth, we need to first examine the pollution within our own minds. We need to uncover the roots of our thinking, the misplaced assumptions and beliefs that have led us down this perilous path of fragmentation. We need to recognize that nature herself is not divided. As Tigua physicist Phillip Duran liked to tell us in the dialogues, there is not one part of nature that is physics "over here," and another part that is biology or biochemistry "over there." Nor is the land divided into "history" and "geography"; the land herself is whole and she is storied.

We have sought to demonstrate throughout this book that the human psyche originates from and is nurtured by the rest of consciousness. Our thinking is never truly apart from nature or society. Thought, language, land, and culture are all interconnected. Because thought does not arise from our individual selves alone, we can never successfully extract psyche from society (as we do in the field we call psychology); nor can we really separate society from psyche (what we call sociology). Psyche and society are intertwined. The same is true for psyche and ecology. Our thoughts are part and parcel of the world we live in.

This is why I believe that the primary purpose of education today ought to be to renew our relationship with the living cosmos. Every student who engages in research should do so for the greater purpose of renewal. In short, it is this magnitude of renewal that I am advocating regarding education—a renewal that emerges from the totality of nature and then flows through how we think, learn, and communicate.

What Is the Greatest Strength and Weakness of Modern Science?

David Bohm once said, "The great strength of science is that it is rooted in actual experience; the great weakness of contemporary science is that it admits only certain types of experience as legitimate."[2] This is a very important statement. A by-product of the rise of the scientific method has been to narrow our field of inquiry. In effect, we have narrowed our way of thinking in an attempt to make the data more reliable. Scientists isolate variables so that they may alter the independent variables and then analyze the results that occur (dependent variables). But if a variable does not lend itself to being measured, it is not considered useful. It is deemed "extraneous" and must be either controlled or dismissed.

The old ways are not concerned with measurement—only experience. This way of being—what Abram calls "listening with animal ears" and "animal eyes"—is both a heightened state of alertness and one that is very open-minded. A wider range of phenomena are felt and experienced, and everything could be important—the way a bug moves, a bird flies, a gust of wind, or a shooting star; all could be significant depending on what else is happening.[3] This is why Tobasonakwut Kinew asks "what kinds of things want to happen together?" He asked because he was already closely observing what was happening.

This way of thinking could be complementary to the scientific mindset, which is also interested in close observation. The difference is that Western science limits what variables are important to observe, whereas ancient (original) thinking would not. Once you adopt an original mindset, certain benefits accrue. It is easier to notice partnership in nature; you see a river or a tree much like a community center, with life constellating around them. You think in terms of ecosystems or watersheds instead of a tree or a river. You become increasingly aware of how things are related rather than how they are separated. And this is why you can ask for and receive help from everyone and everything—as Scarface did in traveling to the land of the Star People—because we are all interconnected.

The ability to see how things relate to each other—particularly in nature—is suppressed in the West. We might conclude that the Western mind works differently, more abstractly, perhaps more precisely, and this might be partly true; but at the same time, this conclusion could be misleading for one simple reason: *all understanding is gained contextually*. We abstract (or attempt to isolate) facts in the West, but "all facts are part of theory," as Alfonso Montuori, chair of the Transformative Studies program at CIIS, likes to say. In short, both modern and ancient understandings are founded in context; the difference is that, in ancient times, the context was the actual place in which we lived—not an abstract theory.

Our lives were once full of meaning because everything was part of a sacred context. The more we can recover this original form of thinking, the better, in my opinion, even if it requires returning to an earlier way of thinking. Like a salmon swimming upstream to spawn, it is sometimes necessary to go backward in order to give birth to new ideas.

Today, we pride ourselves on how far we have come, and in our scientific empiricism—on the precise ways we obtain data. But the original ways of knowing were also empirical, something that is rarely acknowledged. Ancient ways of knowing were eminently practical, based in actual lived experience. The modern scientific method has the benefit of being reproducible, part of a body of knowledge that can, in theory, be applied for future benefit. The ancient ways of learning, on the other hand, had the advantage of emerging from a living origin, which enabled the body to sense something that rational thought cannot.

To give an example of what I am speaking about, let me relate a story Little Bear has told about the time scientists were building dams on the St. James River in Canada on the Cree territory as part of the James Bay Hydroelectric project that was opposed by the tribe, which resulted in the James Bay Northern Quebec agreement, in which the scientists agreed not to build dams in environmentally sensitive areas. One day, the tribal elders objected to a plan of the scientists to build a dam because that particular place was known to be a spawning ground. The scientists were very skeptical, as the place they wanted to build the dam was full of rapids. "This can't be a spawning ground," they scoffed. But the elders politely insisted. And so, with great reluctance, they hauled all their scientific equipment there and proceeded to take various measurements over several days. And to their surprise, they found that the area was indeed a spawning ground, just as the elders said. One elder turns to the other, and says, "Now, it's scientific."

This story is indicative of how we can receive information if we are present in our bodies, open to receiving knowledge from what Indigenous elders refer to simply as "the natures," or knowledge gifted directly from the source. This was the way the ancients learned, again and again. This kind of knowledge or inspiration tends to come when we let go of what we think we know and begin to listen anew.

Contrast this with modern education, where we learn from a textbook or a body of knowledge that is already formulated and stored. A textbook, no matter how well-researched and written, is the definition of inert knowledge. It is knowledge of a dead past that has been separated into categories and stored away, much like we organize a file cabinet. These categories have become the separate subjects of modern academe and are knowledge *about* something. Although this knowledge may be useful, it only teaches *what* to think—not *how* to think.

Can We Receive Knowledge Directly from Nature?

The story modern science wants us to believe is that human history is the accumulation of increasingly correct ideas until we reach the present understanding, which is considered the most correct. And so we falsely project our modern ways of knowing back upon ancient societies; we imagine our forefathers learned everything by trial and error. The ramifications of this belief are manifold. The most debilitating consequence for education is that we deaden inspiration by requiring students to memorize mountains of accumulated information. When students are buried in textbooks, they do not think creatively. They also rarely venture outdoors, where they could be absorbing living nature.

It is possible to receive knowledge directly from nature, and it happens all the time. A writer knows this, and when stuck, goes out for a walk. My first teacher other than my parents was a tree—a wise weeping willow tree that lived on the kindergarten grounds. I used to sit underneath that tree for long periods, basically whenever I could. I vividly remember that tree and her gentle branches that enfolded me like arms, and the soft breeze that wafted through her leaves, whispering me wise and peaceful thoughts. I cannot remember any human school teachers from that time. The rest of kindergarten was rather unremarkable; I played with children I no longer remember, but I still remember that tree.

Why is it that we now believe that we cannot hear messages from nature when, for millennia, we believed otherwise? We make exceptions for our historic religious icons—Moses, Jesus, Mohammed, and the like—who are seen as *exceptional* precisely because they were able to receive the direct voice of nature. We do not call it "nature," of course; we call it "God." I am not saying it is not God, but it is still nature and immanently real, if we, who are of this world, can hear it. The voice of nature or the voice of God is *revelation,* which sounds very mystical, but only means something that was revealed instead of acquired.

Our preeminent American philosophers, Emerson and Thoreau, learned directly from nature. We misunderstood them, calling them transcendentalists and loners when they were anything but. Emerson and

Thoreau were not transcending; they were embedded in the community of nature, especially Thoreau. It was everyone else that was transcending nature—with some exceptions, such as contemporary Indigenous peoples, who had the temerity to do the same thing as our religious icons.

The Necessity of Wholeness in Education: If Not Now, When?

I firmly believe that it is possible to unite the wisdom of ancient interconnection with its modern forms of expression, bringing them together in one university that embodies wholeness. Such a university would cultivate dialogue-based and question-driven education. It would permit both students and faculty to approach the curriculum as one whole without separate departments of specialization. It would go beyond interdisciplinary research to foster transdisciplinary research, so students could follow their inquiry across the boundaries of disciplines, understanding that any one field can be informed by all other fields.

Most importantly, research would be undertaken for the purpose of renewal. This is very different from research conducted for the purpose of accumulating knowledge, which often results in attempts to manipulate and control one's environment from a supposed objective distance. The latter is a remnant of the Enlightenment mentality wherein human beings are imagined as standing outside their environment, and the rational is elevated above the physical.

A truly wholistic university would balance the rational and the intuitive. Creative inspiration and the expression of beauty would be encouraged and understood as necessary for healing and renewal. The overall effect of such a genuinely wholistic and open-minded curriculum would be the emergence of not only creative and original thought, but profound healing and regeneration. This kind of school is not only possible; it is urgent and necessary.

Now, I realize that some readers may think a truly wholistic school is too far from present-day reality and therefore unattainable. But in our rush to validate our modern approach to knowledge, we have forgotten how long education mirrored the principles of the ancient world. This is why I took the time to sketch the history of modern education—to show that, in the larger scope of history, wholistic thinking has predominated.

In fact, because educational practices tend to lag behind the shifts in the larger culture, it is only in the last two hundred years that we have actively embraced fragmentation over wholeness and parted ways with both ancient wisdom and leading-edge science. Because this trend is so recent, I believe there is hope that it can be reversed, or at least counterbalanced.

It is possible to incorporate ancient wisdom alongside contemporary thought in an original and fresh combination. To do so, however, requires a new set of working assumptions regarding humanity, nature, and education. I put these assumptions to paper below with a caveat that they should not be construed as ideology or dogma. They are only rough guidelines because the most important aspect of designing an original model of education is flexibility and openness to Spirit.

In turn, these assumptions may include certain ideas that may appear contradictory or ambiguous. If I am criticized for this, so be it, because original thinking contains ambiguity and paradox—not as opposites, but as complementary poles; otherwise, it cannot be inclusive and most certainly cannot be whole.

New (Original) Assumptions about Humanity

- The human being is inextricably connected with all of nature
- To be human is to seek our purpose within the greater whole
- Each human being is unique, but not of wholly separate consciousness
- Our thoughts are all interconnected; to be is to be related
- An aspect of being human is to think rationally and abstractly, but this is not tantamount to being human
- A human being also thinks with his/her heart, gut, and whole body

- Human actions are part of the larger natural cycles of birth, growth, maturity, and decay in all of nature
- We do not progress in a line; progress that appears linear is only an aspect of the larger unfolding of nature
- Human progress is an unfolding of latent potential contained in the seed of consciousness (buddha nature/atman/kingdom of God) within; we move from seed to root to bud to fruit like all of nature, with our fruit being wisdom
- Human thought, consciousness, and worldviews are derived from a larger movement of Nature's consciousness
- Human beings have free will, and can choose to live in accord with the movement of nature or resist it
- Time is an aspect of human nature; it is not solely outside of us but also within us

New (Original) Assumptions about Nature

- The cosmos has always been here
- All of nature is alive and animate
- All of nature is interdependent
- Thought coarises with nature
- There is a living order to nature
- Nature is also a turbulent flux of unceasing change, creation, dissolution, and rebirth
- Nature is a self-organizing whole
- Nature is fundamentally creative and self-educating; there is an inherent desire within all creatures to learn, grow, and release knowledge for the benefit of future generations
- The elements of nature are not merely the constituents of life but alive themselves

- The cosmos is kincentric—primarily a system of relationships and alliances rather than a competition for survival
- Nature is an undivided whole; real life is not divisible into separate subjects

New (Original) Assumptions about Education

- Nature is the original teacher
- Education comes from *educare* or "drawing forth"; education is a remembering or recollecting of the wholeness of knowledge already existing in nature and within ourselves
- Education is an initiation into one's life purpose
- All subjects are interconnected
- Any one subject can be informed by all other subjects
- Education should be learner centered and question driven
- The pursuit of the inquiry will take the learner into whatever subjects are necessary
- Knowledge of facts must be placed in a larger context of theory or, ideally, in the actual context (place) of living
- Research is not limited to a preexisting body of inert knowledge; it can and ought to be a renewal of one's ongoing place in nature
- Ecology should not only be the study of or about nature, but be conducted in nature, with human consciousness and action considered as part of the whole
- Collaborative learning and collaborative teaching are in accord with nature
- Learning can come from creative insight or direct knowledge independent of trial and error or the accumulation of past results
- All knowledge can always be reexamined and reevaluated
- History is still reverberating in the present; no event is completely over

How Do We Remake Education to Foster Original Thinking?

We once received our stories from places and, in fact, could only tell a story in a particular place—and this was education. But with the rise of perspective, we became the chronicler of nature, which was seen as outside of us, and this changed everything. The meaning of history changed from a detailed inquiry that had nothing to do with time to accounting for what happened during linear chunks (fragments) of time. But the meaning of education changed also. Education was no longer about learning to coordinate with the cycles of nature as they were happening, but was about passing on the knowledge we had accumulated before. This is the point where education moves from being continually relevant to being continually irrelevant—something we have been struggling with ever since. It is also the time when education went from being an inspirational enterprise, one of original thinking based in knowledge revealing itself to us to something completely different—knowledge being accumulated.

As Krishnamurti noted, our psychological desire to accumulate is the root of division, and the precursor of our modern worldview. Once we saw time as apart from us, we removed ourselves—voluntarily—from the timing of nature and began to accumulate man-made knowledge. As we removed ourselves from nature, nature stopped revealing information to us. We then thought we had to, in Bacon's words, "put Nature on a rack to torture her" in order to discover what used to come easily to us. This was the price we paid for disentangling ourselves from nature.

Original thinking, on the other hand, is a direct transmission of knowledge from living nature. This knowledge need not be acquired as in wresting it away from its source. Our original instructions are always available to us as needed, provided we maintain a humble approach of reciprocity with the natural world. We only need remember how to open ourselves up to this living knowledge. We will then receive it as the blessing that it is.

For education to promote original thinking, it must be similarly alive—not a mere accumulation of (inert) facts. This is not to say that the accumulation of knowledge is worthless; I am not suggesting that we throw away the vaccines for polio or smallpox, for example. But let's remember that the vaccine for polio was invented by rethinking the problem in a fresh (original) way. Prior to the polio vaccine, all vaccines were made from living viruses. Salk introduced a dead virus into the body and fooled the immune system into treating it much the same as a living virus without the negative consequences. His daring treatment was physically based, which is the only way the mass culture could believe in its efficacy, but his approach also exemplified a playful trickery that is not altogether different from the techniques of a shaman making psychic interventions to fool a person into expelling an illness or imbalance. Salk followed his intuition and was heavily criticized. His intuition was later coupled with rational processes to implement the cure. The lesson? A body of knowledge needs to be continually questioned and reassessed. Original thinking allows fresh approaches to situations wherein the status quo is no longer working.

I believe that education should include some dissemination of already existing beliefs, but no belief should be sacrosanct. Students ought to continually reconsider the validity and scope of any and all beliefs. Modern education rarely encourages this. Instead, students are programmed to value the accumulation of knowledge. Mainstream ideas are easily accepted, and so they harden into dogma. Genuinely original thinking is met with scrutiny. Of course, there is lip service to the concept of original research in academe—but this research is essentially an appendage; doctoral students are encouraged to "replicate and extend" existing research. Within a worldview of linear human progress, this is the only conceivable pathway.

Modernity is viewed as the beneficiary of centuries of accumulated knowledge, and so students are not expected to think for themselves but to learn what has already been formulated. The function of education—particularly childhood education—has become to deposit preexisting knowledge into the minds of young children.

Education should not be merely for the purpose of indoctrinating the next generation. This approach formalizes education as a repository of accumulated knowledge at the expense of creative and original thinking. Our educational system as a whole is an aggregate of inert knowledge that has been accumulated over time in fragmented bits and bytes. We like to think that these smaller bits of information make up a larger, finite whole, but they really do not. As we have seen, related parts can make up wholes—but fragments cannot. When disciplines and professions cordon off their own territory, ignoring the myriad of connections with other ways of knowing, fragmentation is always the result. And what is worse: a fragmented worldview is insular; it tends to not know that it is fragmented. It cannot know because it has been cut off from its roots. This causes irreparable damage, not to the discipline, but to the society, as we saw in our review of economics.

Complex problems, such as global climate change, world economy, food security, and ecological sustainability, do not conform to disciplinary or national borders. Students who bring original thinking to these issues will serve a vital purpose in modern society; they will be the change agents who can introduce whole and balanced solutions where today there are only fragmentary and partial approaches.

In truth, no subject is an island to itself; all subjects (and all partial formulations) are abstracted from the collective fabric of nature. We cannot piece these fragments together, but we can do something else, which is to dive into the collective soil as a whole. All subjects are interconnected in complex, nonlinear ways, much like a mushroom rhizome. These living, interconnected roots, when accessed, tap into the moving energy of life on Earth.

Chapter 19

A Vision for Original Thinking

At the turn of the millennium, I received a vision. I saw a village of five interconnected buildings. There were four rectangular buildings, one in each of the four cardinal directions, and a fifth, circular structure, in the middle. The central building seemed to be the spiritual source of energy for the village, and it appeared in the vision as a kiva (the Hopi word for a partly underground circular gathering place used for ceremony and spiritual purification). Each of the exterior buildings had entrance and exit doors that served to channel energy in a circle: from east-to-south-to-west to north. But there were also pathways to the center that continually fed the overall energy in complex feedback loops.

At the time, I believed this vision represented a school campus of interconnected studies, and I feverishly wrote down what I saw. I intuited that the center building represented the spirit of the village, with the energy emerging directly out of the soil, and the four outer buildings represented not only the four directions, but also the four elements: fire, air, water, and earth, which continually interconnect and interpenetrate each other. (See below.)

Original Illustration: Dyanne Strongbow

Figure 7: Vision for school campus

I remember being aware that this vision was a large one—and that it would probably take some time to fully manifest. I was wary of trying to understand it too soon—of trying to convert the intuitive into the rational with undue haste.

It seemed clear to me that this was a grassroots model for higher education, and over the next decade plus I stayed faithful to this vision and called upon others to help manifest a model of higher education that was truly a synergistic, interconnected whole.

The group I was working with came to assign the four outer buildings to the four directions and the four elements per the following: science and cosmology (fire, east); ecology (air, south); expressive arts (water, west); and integral healing (earth, north). Original thinking would be at the center (the fifth element of spirit or time) and serve as the catalyst for interconnecting the other elements. The degree itself could be called *Original Thinking*: Science and Cosmology; *Original Thinking*: Ecology; and so forth. I fully articulated the potential program in a doctoral dissertation, which is available online for free under the title: *SEED Graduate Institute: An Original Model of Transdisciplinary Education Informed by Indigenous Ways of Knowing and Dialogue.* An abbreviated form of this vision for higher education is also available on my website: www.originalthinking.us

Is Education Limited to School?

Since that time, my vision of education has continued to develop and change. Although I am still passionately interested in reforming formal education and there are clear benefits that an accredited degree can provide in society, I have increasingly come to see education as something that happens mostly outside of what we call school. My thinking changed in part because of continuing experiences with dialogues. But it also changed for a different reason. I devoted more and more time to being outdoors, in nature, away from other people, actively seeking development of the vision.

As I neared completion of this book, I felt an increasing need to let go of what I thought I had learned—to "empty my cup," as Leroy would say, so that genuinely new knowledge might enter. I had embarked upon a similar type of process before writing the book, but now I had again become too full (if not too full of myself). I began to contemplate doing a formal vision quest or "cry to Spirit" for a new vision.

A caveat to readers: I am not saying that one must embark upon a formal quest to receive a vision of what one needs to do in life. And I am certainly not prescribing it for anyone else. A quest can be an incredibly arduous event. In many traditions, a person will go up to four days in isolation (from other people) with no food or water and little or no shelter from the elements. There is a reason why vision quests are traditionally something that adolescents engage in as a rite of passage into adulthood; young people are clearly better able to withstand the physical demands. But, ultimately, as my friend and vision quest leader Mike Two Bears likes to say: a vision quest is not about suffering; it is about receiving a vision.

I have yet to endure a vision quest of three or four days, but I have completed quests of one and two days, and I know from my own experience that these shorter quests can still produce visions. In my case, I received benefits even before my first quest began—from the moment I contemplated doing it. That night, I had multiple dreams of dying and being reborn into a new awareness, of connecting to ancestor spirits emerging from the ocean and floating through the air. I also began to hear differently, both in my dreams and in waking life. Common sounds—bird calls, branches swaying, leaves whispering—became more prominent. Spirit seemed to be telling me that it was all about listening. I could feel the muscles behind my ear twitching in anticipation, and I eventually realized that I was being prepared for something that my rational mind could not possibly understand.

When the time came, I set out on the quest, high in the mountains above Cloudcroft, New Mexico. I selected and set up my space as I was instructed to do, drawing my circle with my prayer stick and placing four rocks at the four cardinal directions and my altar in the center. I had also brought sacred objects of my own, including sage, tobacco, and bear and eagle fetishes. I brought Tibetan bells and a singing bowl that I have used for years—or should I say for *the ears*—since they serve to awaken one to hearing increasingly subtle vibrations. I knew the bells would be important. I blessed the space with tobacco and said my early morning prayers, chanted, prayed more with the bells and singing bowl, and then sat down to listen to the sounds around me.

Almost immediately, I experienced an acute increase in my hearing that was rather astonishing. I heard ravens from a great distance speaking in their distinctive staccato voice—CAH CAH. When the ravens came closer, I heard them make guttural, rumbling sounds that I had never heard before. Then, a blue jay spoke in an even sharper tone, almost hawk like, and at a much higher pitch, JAAY, JAAY. The bird was far away but eventually came into my field of vision, and flew directly toward me, coming to rest about fifty yards away and remained silent for a time. It was then that he opened his wings, and I was surprised to hear the sound—and loudly. In fact, the sound of the wings was not "out there," but in my body. I felt waves of electric vibration coursing through me, opening me to a frequency I did not know existed. The jay then came up to me, sitting on the ground beside me, clearly acting as a welcoming ambassador. We exchanged something, but I cannot translate the language, at least not yet.

A particularly magical time was when a honey bee came to linger at several flowers that were in my circle. I experienced the relationship between the bee and the flower as the lovemaking that it is. While the bee was still there, a soft rain began to fall, and then a more insistent downpour. I could feel Mother Earth opening up to receive the blessing from the sky. This too was a form of lovemaking, with Father Sky impregnating the earth with future growth. After the rain tapered off and then ended, the blue gramma grass, still twitching in ecstasy, began dancing with the breeze in what seemed to me a conscious celebration. The sun shone upon its body, and the raindrops glistened. A great peace enveloped me.

Gradually, I slipped out of rational mind, as if I were taking off excess clothing. Instead of focusing my egoic attention outward, I allowed events to present themselves to me and to become truly wrapped up in them. Again, it was sound that was the most prominent shift. I realized I had entered an entirely different world—one in which sound was no longer the operative word. It was a world of vibration. And everything that exists vibrates. This world seemed to have infinite layers upon layers of subtlety, and I was only able to get a glimpse of it all—to at least realize that there were vast dimensions I could not see.

As the afternoon shadows darkened, I marveled at the perfectness of nature's timing. So much of human action is about trying to control what is, to know "what time it is," to select and separate one event to focus upon over another. But nature is not a sequence of events; it is an orchestra of things happening all at the same time. A vision quest or other forms of sacred contemplation can teach us to soften our focus, to accept things as they are without trying to control them with our minds. It is all about trust.

Much of what happens on a vision quest—or any time in which rational thought is suspended—is hard to describe in words. Rational thinking prevents us from becoming involved in nature. But when we do become involved, we sense something of the unlimited reservoir of blessing and knowledge that has been available to us all along. On my quest, I came to see every blade of grass, every plant, rock, tree, or insect as a teacher. I felt their energy as aware and alive and imbued with spirit. I knew that Spirit was all around me and moving though me, carried by what I used to call "wind."

At times, it was blissful, like being hugged by God. At times, it was intense and challenging, because of the sheer magnitude of the power I was experiencing. In the end, I came away with a vision that did not replace my vision of reforming education but deepened it. I now realized that we need to reeducate our minds even prior to reforming education, and that the reform of education will naturally follow when our minds are ready.

I have sensed this before and thought I knew this before, but I did not really know it. Through vision quest, I experienced a deepening of something that my rational mind knew about like a traveler might know about a distant land they had not yet visited through a travel brochure or movie. On my first quest, I went into it being intellectually aware that we are all composed of 70 percent water. But it was quite another thing to go without water and to feel right down to the marrow of my bones that: I AM WATER and from that place to have a new relationship with all the waters in the world. Direct knowing is different. It is an experience, not an abstract learning about something.

What Are the Ramifications of Firsthand Experience for Education?

My own experience has taught me that education is something that is always occurring in nature—and always occurring within myself—unfolding at its own pace. Our task is to tune into what is already ongoing. This kind of education happens all the time outside what we know as school, but can also happen inside school.

Formal education, even at the level of higher education, could be restructured to allow more of a balance between natural experience and a rational interpretation of that experience. Although it is true that there is something lost when experience is put into words, it can also be enormously helpful to give voice to one's experience. For this, dialogue can be a bridge. This is because dialogue, while ostensibly done through words, is also an experience of listening to Spirit. Even when speaking, the primary experience ought to be listening, allowing the words to speak through you in the same manner as in saying a prayer.

In a good dialogue, participants learn a lot, but only after *unlearning* what they think they know. Little Bear playfully calls this "brainwashing"—to literally wash the brain of all that it has accumulated and clings to as knowledge. This happens naturally—and sometimes dramatically—during the course of a well-moderated dialogue. The moderator throws out a kick-start question or statement that is often a koan—and locks up the brain—disabling its normal patterns of thought so that new thoughts have a chance to enter. The questions we have asked in this book: is it possible to have an original thought; what does it mean to be human; and so forth, were similarly intended. To consider them properly, as a writer or reader, one has to set aside, if not drop, one's preconceived notions.

In this book, I have sought to approximate the process of dialogue, allowing meaning to deepen over time. A similar thing occurs in an oral dialogue; the process is slow, and for some people, excruciatingly so. I know it took me some time to get used to the process. I had a breakthrough when I considered the origins of dialogue, which hearken back to the origins of humanity, listening to stories told around a fire. When I consider why these origin stories are so powerful, I realize it is because they are timeless; they are as alive now as they were eons ago. Why is that? I believe it is because these stories, like dialogues, dreams, and visions, are infinitely rich and multidimensional. They are riddled with meaning.

The ancient Greeks understood that riddles were the most effective way to educate. But we moderns have grown impatient, looking for a moral to the story—something to tie it all up neatly in a bow. We have traded stories for information. Information is single-pointed, but an origin story is multifaceted, with layers upon layers of meaning. These layers of meaning unfold over time, and have a way of coming to the surface just when they are needed. A mysterious synergy of hidden forces always leads us to our destiny. In ancient times, we called these forces "the fates," three mythological goddesses who were originally sisters and the keepers of destiny. In Norse mythology, the Sisters of Fate appear at the center of everything and represent the entire flow of time and the ongoing renewal of the cosmos.[1]

Story enables us to work through our own sense of purpose and destiny, to discover our fate. Fate is also something that restricts—that takes us back to the living story we are called to and points us in a particular direction. We may deny our fate; many of us do. We moderns do not like to be fenced in. But each of us has a destiny—a story waiting to be told. As long as that story is not nipped in the bud by internalized self-conditioning coming from parents or society, it will reveal itself in due course. And when a person finally resolves to live the story they are meant to live, everything changes. All of the forces, elementals, and muses come out to play; they join in support of the right path. The universe is a friendly and supportive place if we have the courage to give up what we think we know in exchange for an unknown future.

So What Exactly Is Original Thinking?

Until now, I purposely refrained from formulating a single definition of original thinking as I did not want a definition to get in the way of an unfolding process. And even now, I will give a definition only in the context of 13 aspects or qualities, not one. This is because original thinking is not any one thing. In truth, original thinking is more of a process than a thing, more of a verb than a noun. In my opinion, it can never be defined completely. With that caveat, below is a graphic of the thirteen qualities of original thinking that emerged over the course of this book, followed by a brief summation of each characteristic. The thirteenth quality: wisdom, will be addressed at the conclusion of the book in Final Thoughts.

Thirteen Qualities of Original Thinking

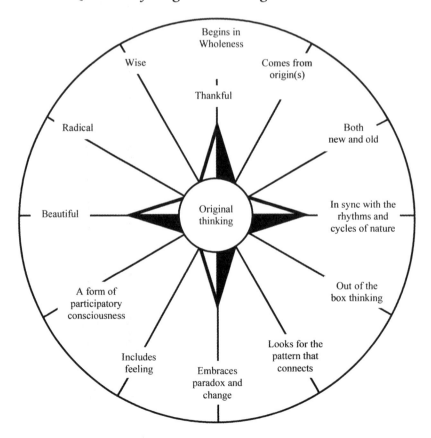

1. Original Thinking is thankful.

> Originally, thinking was a prayer that came from a place of
> blessing, recognizing the already existing completeness of life.
> At its root, thinking is thanking.

2. Original thinking emerges out of wholeness

> Original thinking is a direct transmission of Spirit that begins
> in wholeness and connects us with the whole of creation. Our
> consciousness is an inlet to a larger fountain of wisdom, and
> we are inextricably connected to that greater source at all times
> whether we realize it or not. Grandfather Leon referred to that
> wisdom as coming "from the cosmos into the core, the center
> of Mother Earth," and from there, it comes up from the place
> we are in, and we receive it in our individual way in keeping
> with our unique signature vibration.

3. Original thinking comes from particular places (origins)

> Original thinking comes from particular places that con-
> tain particular vibratory frequencies and inspire particular
> thoughts. If we attune to a place, we become aware of a deep
> knowing that is emergent from the place we are in. Our
> thoughts come from the place we are in whether we are aware
> of it or not.

4. Original thinking is not new or old but both

> Once we become aware that our thoughts begin in wholeness
> and are an inlet to the whole of creation, we can also sense the
> continuing vibration of what is normally referred to as "the
> past" and how it affects what we call "the present." Ultimately,
> this new awareness allows us to become an open doorway
> between past, present, and future. We are able to braid the past
> and future together in an alive presence of the moment. We
> may receive prophetic visions that go beyond our normal con-
> structs and constrictions of the rational linear mind. All of us
> have an opportunity to experience this in our dream life, which

is a good place to begin to observe nonlinear consciousness. But eventually we may be able to see how this wider form of consciousness bleeds into our waking life.

5. Original thinking is in sync with the rhythms and cycles of nature

There is a pulse and rhythm to nature, and we humans are capable of synchronizing with it because human beings also progress in cycles, even if we are not aware of it. A tree goes from seed to root to bud to fruit and then back to seed in order to start the whole process again. Human beings progress in a similar way, even as we have convinced ourselves otherwise. For millennia, human traditions and ceremonies were based on the awareness of life as a circle, and among intact traditions, this is still done. The circle of life helps us act in harmony with all things as they arise and fall in their proper season.

6. Original thinking is out of the box thinking

Linear thinking sees reality as partitioned into separate things, categories, and time blocks. Original thinking is outside the box of linear fragmentation, seeing beyond and across separations, categories, and disciplines. It is wholistic, meta-paradigmatic and timeless, connecting old and new in original and creative combinations of thought.

7. Original thinking looks for the pattern that connects

In the spirit of the ancient natural philosophers and moderns such as Gregory Bateson, original thinking looks for the pattern that connects. It sees the world as all related, as "kincentric." This approach naturally lends itself to seeing partnerships and alliances in nature rather than separation and competition. The continued observation of partnerships expands thinking in increasingly inclusive, interconnected, and whole ways. It is an ongoing process—never completed but ever widening in scope.

8. Original thinking embraces paradox and change

Original thinking is both/and rather than either/or thinking.
What appears to be a polarized end of a dichotomy is really a
complementary balancing side and this can change over time.
As Grandfather Leon said, if given enough time, "a negative
becomes a positive and a positive becomes a negative."

9. Original thinking includes feeling, instinct, and emotion

Original thinking is a total movement of thought that originates in nature and enters the body through the feet, gut, or
heart, and only later is understood by the head. To think originally is to accept instinct, emotion, and feeling as part of the
continuum of thought.

10. Original thinking is a form of participatory consciousness

Original thinking revives the old meaning of consciousness,
which is "thinking with" rather than in separation. It is a form
of participatory consciousness or thinking together that is very
creative. It is not "group think" as in mindlessly following, but
an active form of group knowledge construction and collaborative problem solving.

11. Original thinking is beautiful

Original thinking is beautiful because it brings human thought
in harmony and balance with the thought of all creation. Ironically, this form of beautiful thinking is similar to the roots of
rational thought—rational coming from *ratio*—or the relationship between things. In the right proportions (what the
ancient Greeks called divine proportion or the golden ratio),
the relationship between things is beautiful. But there is also
something very immediate about our ability to perceive beauty,
something unadulterated by rational analysis. When beauty is
first felt, it is direct, immediate, and awe inspiring. Ultimately,
beauty can be a path of life, what in Navajo is called *sa'ah*

naaghai bik'eh hozhoon, or the Beauty way. Living in beauty begins with and is nourished by beautiful and loving thoughts.

12. Original thinking is radical

Original thinking is radical in the original sense of going to the roots. It is about asking the right questions to get to the bottom of things and is of great depth and dimension. Original thinking is analogous to the way the Dogon people construct a basket. Dogon baskets have a rectangular base and pyramidal top. The name for the rectangular base is the word for earth or matter—what is manifest—and the name for the pyramidal top is the word for sky, heaven, or what is unmanifest but where all possibilities reside. The pyramid is placed on top of the rectangular base and weaved together with a thread.

13. Original thinking is wise

Summary

Throughout this book, we have sought a deeper way of thinking that we have called original—or thinking that is not new or old but comes from a place of timeless origin—a place of relationships and alliances that is inclusive, interconnected, whole, and wise.

We began by examining the unconscious assumptions that inadvertently divide our consciousness from its source—assumptions about time, space, place, and origin. We explored the gradual shift from original participation in nature's cycles of change—what I would call "real time," or real timing, emergent from one's connection to place—to the modern notion of abstract time, something that happens apart from our direct involvement. This modern concept of time is of a separate dimension that we then further separate into quantifiable intervals. It is independent from us and from the rhythms of any particular place.

We went on to explore how this divorce from place and the timing of nature brought about a concomitant split in our thinking. Our consciousness, once understood as derived from relationship or "knowing with" other creatures who coinhabited our place, came to be seen as individually created and something only humans possess. We retreated into our heads and saw the world from our distinct vantage point rather than being integrally involved with the rest of nature. As we chronicled, much of this change became reified in the Renaissance with the invention of linear perspective in art. "Realism" in art became our worldview and, by extension, the way we defined ourselves as human beings. Man became known as the "rational animal," the one who is not only separate from nature—viewing it from a detached perspective—but proud of it.

Our divided consciousness, eye separate from nature, ego separate from world is, in truth, a perception or representation of the real. It is not reality itself, but we mistakenly believe that it is. This is what Whitehead called "the fallacy of misplaced concreteness,"[2] and we saw how it pervades the modern educational system and the professions that come out of it. The way that we think has profound effects upon society; in fact, it creates society. Society is fragmented only because we think in fragmented ways.

We have perpetuated fragmentation by dividing nature into abstract segments, studying each subject as if it were reality in and of itself rather than a representation of reality, and a partial one at that. We further divide each subject into bits of inert knowledge that we proceed to build upon (accumulate) in isolation from all other subjects. We ignore the underlying connection between aspects of nature. We do this almost unconsciously, because our tacit infrastructure, infused with a linear perspective, perceives reality as something out there, apart from ourselves, a puzzle of individual parts rather than an interrelated whole.

The ramifications and consequences of our modern worldview are potentially cataclysmic. But there is always the possibility for radical reform and change in our thinking. To reform our thinking, we have to first realize how it arises from nature, moment to moment, within each individual, and how it spreads to the collective society, in drops, rivers, and oceans of thought. It is for that reason that much of Part Three (Return) focused on the overall process of thought from origin to manifestation.

In this last part of the book (Renewal), we reimagined education from the ground up, remembering a time when education existed before there were schools, and its purpose was to understand our place in the world and what it means to be human. Part Four had a dual purpose of reimagining both formal education and the reeducation of our minds to receive inspiration directly from nature. Storytelling and dialogue are also portals to original thinking. All are for the purpose of renewal.

The concept of education for renewal—education that reinforces our connection with Spirit and with all that share existence with us—is a beautiful thing, but not something I invented or a new idea at all. I learned the concept from Blackfoot elder Little Bear who, in turn, is echoing the concern of elders and wisdom keepers from many diverse traditions across eons of time.

We have now moved through four distinct kick-start questions that are also very much interrelated. The questions about original thought; what it means to be human; how we create the world with our thinking; and the renewal of original thinking through education are conjoined by a common purpose: that of remembering our interconnections with our place of origin and all our relations. The modern way of thinking has sadly and systematically deracinated us from our roots. The way we think today about time, progress, humanity, education, and nature are all based in abstract disconnection.

In this book we have taken a different path by braiding together thought, time, humanity, nature, and education as one interconnected whole. The sum total amounted to a reeducation of my own mind and hopefully that of the readers. I took this journey in homage to the ancient natural philosophers and moderns like Bateson who sought out the patterns that connect the whole. It was never a mere philosophical exercise or exercise in nostalgia. I have sought to recover a small piece of wisdom for a specific intent—that I may join with others in helping to restore the collective wisdom of people and nature into all aspects of society. This cause is increasingly popular and increasingly urgent, for we are at a special time in history, a time that is both dangerous and of great opportunity. The danger is that we will perish from the earth, but the opportunity is for us to restore and reintegrate wisdom—now and for future generations. If we do so, we will once again live in harmony with our fellow human beings and with all our relations in which we share this planet. I pray that this will be so.

Final Thoughts

Original thinking is a resurgence of something that has been forgotten for so long that it is considered new again, as in Abram's story of the King and the map. We have gone on a journey, a journey we have come to believe is truth, but is fantasy, a make-believe world of the mind separated from the earth, separated from our place of origin, the source of our own consciousness. We live so apart from nature that, when she peeks through our veil of abstraction, we are indeed surprised, unsure of just what is this old stuff that is resurfacing. But it is resurfacing nonetheless, and eventually it will penetrate our awareness.

The map is our rationality; it must be pierced in order to unearth the layers of wisdom that lie below. This wisdom still exists all over the planet in pockets of societies that remember the old ways—teachings that once flourished and can do so again. These ways are not abstract, but real. They directly engage the human species with what is other than human, opening us to learn from all there is. And there still is so much to learn. We may never learn how to build a ceramic as durable as an abalone shell,[3] or a thread as flexibly strong as spider silk. We have not even mastered dams as efficiently as a beaver. But just the simple act of noticing the wisdom of nature—and not dismissing anything we do not understand as mere "instinct"—is inspiring. The more we learn from others, the closer we come to understanding our original purpose here on Earth.

None of these original instructions are instructions in the way we have come to think of the term. They are not a handbook of information or a precise step-by-step guide. Wisdom does not work that way. Wisdom is born of real experience, but not in the way we commonly assume. It is not garnered through the accumulation of past results, for if it could be, then computers would be wise. Wisdom, as a friend once said to me, "cannot be googled."[4] Wisdom is ineffable. It may be communicated, and often is communicated, without words. It may come though vision, dream, story, a bird song, or a flash of lightning. But it will not be received if the mind is not receptive.

The people I consider wise are very present in the moment and have undoubtedly been so for some time. They become wise through an attitude of open engagement. It is not the accumulation of life experience that makes them wise, but the way they have experienced life. They know things in a way that many of us do not simply because they trust in their intuition. They may have surprisingly analytical minds when needed, but they do not rely solely upon rational thought. They allow for the mysterious to augment their moment-to-moment awareness—and it is this quality that makes them wise. A wise person has wisdom vibrating through every aspect of their being; you can feel it just by being in their presence.

Today, we talk all the time about information, and we are said to live in an information age. But we rarely talk about wisdom. We need a resurgence of wisdom because information, while itself limitless, does not equate to understanding. What happens when we break down enormous amounts of information into bits and bytes of disconnected pieces? We cannot remember it, and so we become dependent upon computers—which are remarkable tools—to store and retrieve all this information. This enables us to possess, by extension, the same amount of information a computer can store. We become more knowledgeable, but we do not become wiser; if anything, we become less wise and less interested in wisdom because we are so wrapped up in processing mountains of information.

Wisdom promotes understanding between peoples. It comes through a very old form of technology—dialogue—in which minds are merged, not separated. In an early dialogue, Little Bear once referred to this shared meaning space as a "lake of authenticity." At the time, I formed an image of Western and Indigenous peoples talking to each other across the divide of a canyon, but with dual waterfalls on both sides leading to the creation of a common pool beneath. No matter which side of the divide one was on, the water below reflected the truth that was shared.

This pool was calm, unlike the turbulence above. I have held onto that image because it speaks volumes to me.

For humanity to survive it will be necessary to look deep into the well of experience and remember what original thinking and the original experience of being human really is. We can no longer be satisfied with what passes for originality—and by this I mean no disrespect to clever innovations such as an iPad or Kindle—but we must rekindle a different fire. We must rekindle a fire not for better living through technology, but for better living through wisdom; we must rekindle a fire for kindness and compassion through the flowering of the human heart. We must think not with our brains or head alone, but with our feet, gut, heart, and soul.

For humanity to survive, we need not give up all our technologies, but we do need to give up our dependence upon them. Technology can be a useful tool if it is the servant of original inspiration, but that inspiration still comes from nature. If we are using technologies that do not promote a healthy balance of life on Earth, as is the case with many, then we need to pause and reconsider our path. There is still wisdom available to us, more than we realize. The vibration of the ancestors has never gone away. If we open our hearts and listen, it will come.

I ritually practice this every morning. I pray to the ancestors who have meant a lot to me and continue to mean a lot to me. I pray to Grandfather Tobasonakwut Kinew, to Grandfather Leon, to Moonhawk, and to Bohm, Krishnamurti, Paula Gunn Allen, Patricia Clark Smith, my parents Joan and Juan, and many others, including my four-legged ancestors who offered me such great companionship. I thank them for everything they have done and are still doing for me now. These ancestors are not really gone as long as they are blessing me, protecting me, feeding me. Some may doubt that we can speak to our ancestors, but I know we can. I sense an open doorway, a vibrational portal—and saying the prayers keeps the doorway open.

The doorway opens both ways of course. I not only receive wisdom from the ancestors; I am able to give love and gratitude back. Sometimes I sense impatience from the ancestors. It seems they are wondering what is taking me so long. In so many ways, the ancestors inspire me, even prod me, to go on and persevere. As Zoroaster once said, "To the persevering mortals, the Immortals are swift." I have never completely understood what Zoroaster meant by that, but it is one of those koans that makes me contemplate the radical interconnection of life and death, past and present.

An intriguing description of the coexistence of past and present occurred in the 2004 SEED dialogue when Nancy Maryboy spoke about the Navajo word for original awareness. This word is used when communicating with the stars and is very beautiful to hear. It has four syllables: *tsu, tsu, long i* (pronounced like ee), and *yo,* or *tsutsuiyo.* Nancy told us the word means: "my ancient relation from whence I came." In short, when she says *tsutsuiyo,* she is talking with the stars as relatives, maintaining a familial relation with an original place and time.

As it happened, Brian Swimme was in that particular dialogue circle— and Swimme had been one of Nancy's mentors in her study of Western astronomy. In a Western way of thinking, of course, starlight is light from the past; in other words, when we see it, we are seeing something that occurred long ago. But Nancy explained that there was no inconsistency between this view and the Navajo view that welcomed our ancestors into the present time. She told us that "the stars are light, and light is consciousness."

My own way of interpreting what she said is that *tsutsuiyo* is a physical manifestation of the origin of our own consciousness. It is not a metaphor or an abstraction because the past and present are conjoined in that encounter. The light exists in the present moment, regardless of when or where it originated. Original consciousness—or original thought, is ever present.

So, as we bring this book to a close, I invite the reader to close your eyes and imagine for a moment the possibilities of *tsutsuiyo*—the original light of our own consciousness. As you breathe in, imagine that your breath is coming from far away, from every point in the cosmos. There is no difference between the breath inside you and the breath outside you. It is one breath. And it is effortless. Give thanks for the gift of breath, and remember to thank the trees who convert it into the form we need as they are already thanking you in return. And thank the dark soil of Mother Earth without which neither you nor the trees could live. And thank the water that is within you, the trees and every other creature. Thank the water that is all over the earth, in the oceans, in the sky, in the mountains, rivers, and lakes.

Next, go within and feel the light energy that is in you, emanating from your solar plexus. It is here where you are connected to your human birth mother, and it is here where you are connected to the entire cosmos. Feel the light within you that gives you life and energy and give thanks to the closest star, our Sun, for it keeps you alive.

Now, use your mind to travel to the place of the Star people. Go up, up, beyond the atmosphere. Look back on the atmosphere of the Earth. Notice the similarity between the way lightning flashes in the atmosphere and the way the neurons of your brain fire. Your brain is part of a larger brain. Give thanks for that. Now, go further out into space and see the perfectly formed orb of our planet Earth. How beautiful she is. Give thanks to her.

Next, I invite you to continue your journey. Imagine traveling farther in space, past the Sun and out into the galaxy. Expand your thought outward and continue your journey by using your mind. Go to the farthest star you can see. Travel through the entire galaxy to get there. Now, if you dare, go through that dark portal until you are outside the galaxy. Turn toward the great expanse beyond—the limitless cosmos.

Then, look back at the Milky Way and see it from the outside in. See all the interconnected points of light throughout the galaxy. You have reached the origin of your consciousness, the vast emptiness that is the original seed of everything. Stay there for as long as you like. Do not think. Do not do anything. You can come back to this place anytime you want. But for now, return to Earth because you are about to have an original thought.

NOTES

Preface

1. For more on David Bohm's philosophy of dialogue, see Bohm, David. (1996). *On Dialogue* (Lee Nichol, Ed.). New York, NY: Routledge.

Introduction

1. The proto-Germanic *pankaz* means "thought" and "gratitude" and is related to the Proto-European *tong,* meaning to "think" and "feel."

2. Bohm, David. (1996). *On Dialogue* (Lee Nichol, Ed.). New York, NY: Routledge, pp. 85–95.

Chapter 1: Original Thought, Time, and the Unfolding of Consciousness

1. A similar observation was made at the dialogue itself by M. J. Zimmerman.

2. Cajete, Gregory A. (1994). *Look to the Mountain: An Ecology of Indigenous Education.* Durango, CO: Kivaki Press, p. 87. Cajete observed that the etymology of the word *indigenous* comes from the Latin *endo* or *indo,* which is related to the Greek *endina,* meaning literally the "guts" or "entrails" of the Earth.

3. Bateson, Gregory. (1972). Form, Substance, and Difference. In *Steps to an Ecology of Mind* (pp. 454–471). Chicago, IL: University of Chicago Press. I first heard of this phrase from Lee Nichol quoting Bateson at the SEED dialogues. Bateson uses the phrase to express what constitutes "information." The full essay discusses how the mental world of information is not limited by the skin but extends out into what we normally call the exterior world.

4. *The Ever-Present Origin* is the English title of the book by Jean Gebser originally written in German as *Ursprung und Gegenwart.*

5. David Abram made this point at several of the SEED dialogues.

6. The phrase "mythical consciousness" is frequently associated with Mircea Eliade, along with "sacred" and "profane" time. For more, see Eliade, Mircea. (1957). *The Sacred and the Profane* (Willard R. Trask, Trans.). New York, NY: Harper & Row.

7. Barfield, Owen. (1967). *History in English Words*. Gt. Barrington, MA: Lindisfarne, pp. 105 and 167.

8. Barfield, Owen. (1967). *History in English Words*. Gt. Barrington, MA: Lindisfarne, p. 167.

9. de Snatillana, Georgio, and Von Dechend Nonpareil, Hertha. (1977). *Hamlet's Mill*. Boston, MA: Gambit, p. 333.

10. Kant, Immanuel. (1997). *Critique of Pure Reason* (P. Guyer, and A. Wood, Trans.). Cambridge: Cambridge University Press.

11. Gary Snyder in Abram, David. (1996). *The Spell of the Sensuous*. New York, NY: Vintage Books, p. 173.

12. Gebser, Jean. (1985). *The Ever-Present Origin* (N. Barstad with Algis Mickunas, Trans.). Athens, OH: Ohio University Press, p. 168 (originally published in 1949 as *Ursprung und Gegenwart*).

    In mythology, Kronos was one of the Titans (elder Gods before the Olympians), son of Uranus and Gaia, and father of Zeus and Hera, as well as Demeter, Poseidon, and Charon. In an attempt to undo the prophecy that one of his children will gain power over him, Krosnos devoured all his children but is forced to disgorge them later and banished by Zeus. Gebser takes the myth to mean the triumph of wakefulness and the birth of time (temporality).

13. Little Bear, Leroy. (2000). Foreword. In Gregory Cajete, *Native Science: Natural Laws of Interdependence*. Santa Fe, NM: Clear Light Publishers, p. xi.

14. Ross, Rupert. (1992). *Dancing with a Ghost: Exploring Indian Reality*. Markham, Ontario: Reed Books.

15. Peat, F. David. (1996). *Blackfoot Physics: A Journey into the Native American Universe*. New York, NY: Fourth Estate, pp. 33–38.

16. Little Bear talks about this concept in the film *Journeying to Turtle Island* by Miryam Servet.

17. Aristotle, http://dupagetherapy.com/Aristotle_On_Imagination.html.

18. This is something Leroy Little Bear likes to emphasize.

19. *The Art of War* from the sixth century BC.

20. See Dewey, John. (1938). *Experience and Education.* New York, NY: Touchstone. In it, he demonstrates how neither "traditional" nor "progressive" education is adequate and advocates instead for a comprehensive philosophy of education.

21. Senge, Peter; Sharmer, Otto; Jaworski, Joseph; and Flowers, Betty Sue. (2004). *Presence: Human Purpose and the Field of the Future.* Cambridge: Society for Organizational Learning, pp. 209–215.

Chapter 2: Looking Backward to Go Forward

1. Barfield, Owen. (1967). *History in English Words.* Gt. Barrington, MA: Lindisfarne.

2. Gebser, Jean. (1985). *The Ever-Present Origin* (N. Barstad with Algis Mickunas, Trans.). Athens, OH: Ohio University Press.

3. Alford, Dan Moonhawk. Manifesting Worldviews in Language. In *So What, Now What: The Anthropology of Consciousness Responds to a World in Crisis* (Matthew Bronson and Tina Fields, Eds.). Newcastle: Cambridge Scholars Publishing, p. 297.

4. Gebser, Jean. (1985). *The Ever-Present Origin* (N. Barstad with Algis Mickunas, Trans.). Athens, OH: Ohio University Press.

5. Alford, Dan Moonhawk. Manifesting Worldviews in Language. In *So What, Now What: The Anthropology of Consciousness Responds to a World in Crisis* (Matthew Bronson and Tina Fields, Eds.). Newcastle: Cambridge Scholars Publishing, p. 298.

6. Whitehead, Alfred North. (1925). *Science and the Modern World.* New York, NY: Free Press.

7. This was something Moonhawk said in an interview with Harlan McKosato after the first SEED dialogue in 1999. McKosato replied: "That's deep."

Chapter 3: Wheels within Wheels

1. This point of the cycle is when spirit first realizes that it has a physical body, analogous to Adam and Eve in the Garden of Eden.

2. I originally heard this saying from my friend Pravin Agarwal, a native of India, who told me this at the California Institute of Integral Studies in the winter of 1983. I find this saying to be very elegant and easy to remember.

3. Blavatsky, Madame Helena P. (1974). *The Secret Doctrine. The Synthesis of Science, Religion, and Philosophy.* Los Angeles, CA: The Theosophy Company.

4. [W]holistic and [w]holism are intentionally spelled with a w—as in "whole"—throughout this book.

5. "Original mind" and "ground of being" are Buddhist terms that refer to the pure, original, or undifferentiated state from which all phenomena arise. While I occasionally use these terms myself, I am not intending to equate "original thinking" with these more specific Buddhist terms.

6. Johnson, Kenneth. (2009). The Shape of History: Time and the Mayan Calendar. *Parabola, 34*(4), 12–19.

7. Calleman, Carl Johan. (2004). *The Mayan Calendar and the Transformation of Consciousness.* Rochester, VT: Bear & Co., p. 5.

8. Calleman, Carl Johan. (2004). *The Mayan Calendar and the Transformation of Consciousness.* Rochester, VT: Bear & Co., pp. 12–20.

9. My thanks to Carole Rominger who reminded me of this in conversation one day.

10. Language of Spirit dialogues, 2001 transcript, Albuquerque, NM. SEED Institute, p. 138.

Chapter 4: It's About Time

1. According to Madame Blavatsky, the ancients had an esoteric word for the animating principle of the cosmos: "Fohat."

2. Begay, David, and Maryboy, Nancy Cottrell. (1999). Nanit`a Sa` ah Naaghai Nanit`a Bik`eh Hozhoon [Living the Order: Dynamic Cosmic Process of Diné Cosmology]. *Dissertation Abstracts International, 60*(05), 1596A (UMI No. 9930321).

3. Gunn Allen, Paula. *The Sacred Hoop: Recovering the Feminine in American Indian Traditions.* Boston, MA: Beacon Press, pp. 11–12.

4. Traditional Blackfoot Story. This version appeared in Cajete, Gregory. (2000). *Native Science: Natural Laws of Interdependence.* Santa Fe, NM: Clear Light, pp. 250–255.

Chapter 5: Purpose, Potential, and Responsibility of Being Human

1. Not his real name.

2. Neidhart, John G. (1932). *Black Elk Speaks: The Life Story of a Holy Man of the Oglala Sioux.* New York, NY: Pocket Books (a Simon and Schuster division of Gulf & Western Corp), pp. 164–165.

3. For more, see: Benyus, Janine. (1997). *Biomimicry: Innovation Inspired by Nature.* New York, NY: Harper Collins.

4. Emerson, Ralph Waldo. (1926). *Essays by Ralph Waldo Emerson.* New York, NY: Thomas & Crowell, pp. 232–233.

5. Abram, David. (2010). *Becoming Animal: An Earthly Cosmology.* New York, NY: Pantheon (division of Random House), p. 193.

6. Darwin wrote about this in a book published after *The Origin of Species.* See Darwin, Charles. (1872). *The Expression of the Emotions in Man and Animals.* London: John Murray.

Chapter 6: Rational Thought and Human Identity

1. Bohm, David. (1980). *Wholeness and the Implicate Order.* London: Routledge, p. 26.

2. McConeghey, Howard. (2003). *Art and Soul.* Putnam: Spring, p. 2.

3. McConeghey, Howard. (2003). *Art and Soul.* Putnam: Spring, p. 1.

4. I credit my friend, Robert Schimelpfenig, a practitioner of Socrates café, for informing, influencing, and inspiring me in various conversations related to Socrates.

5. Ariew and Grene. (1995). *Descartes and His Contemporaries: Meditations, Objections, and Replies.* Chicago, IL: University of Chicago Press, pp. 1–2.

6. Ariew and Grene. (1995). *Descartes and His Contemporaries: Meditations, Objections, and Replies.* Chicago, IL: University of Chicago Press, p. 86.

7. Descartes, Rene. (1951). *Meditations on First Philosophy* (Laurence Lafleur, Trans.). Indianapolis, IN: J. Boss-Marrill Company, Sixth Meditation, p. 82.

8. Descartes, Rene. (1951). *Meditations on First Philosophy* (Laurence Lafleur, Trans.). Indianapolis, IN: J. Boss-Marrill Company, Fourth Meditation, pp. 50–60.

9. Kumar, Satish. (2013). Three Dimensions of Ecology: Soil, Soul, and Society. In *Spiritual Ecology: The Cry of the Earth* (Llewellyn Vaughan-Lee, Ed.). Point Reyes, CA: The Golden Sufi Center, p. 131.

10. Descartes, Rene. (1951). *Meditations on First Philosophy* (Laurence Lafleur, Ed.). Indianapolis, IN: J. Boss-Marrill Company, Second Meditation, p. 24.

11. Lao Tsu. (1997). *Tao te Ching* (Gia-Fu Feng and Jane English, Trans.). New York, NY: Vintage Books (a division of Random House), p. 25.

12. Anthes, Emily. (2013). *Frankenstein's Cat: Cuddling Up to Biotech's Brave New Beasts.* New York, NY: Scientific American.

13. Anthes, Emily. (2013). *Frankenstein's Cat: Cuddling Up to Biotech's Brave New Beasts.* New York, NY: Scientific American, pp. 34–36.

14. Benyus, Janine. (1997). *Biomimicry: Innovation Inspired by Nature.* New York, NY: Harper Collins.

15. Benyus, Janine. (1997). *Biomimicry: Innovation Inspired by Nature.* New York, NY: Harper Collins, pp. 254–278.

16. McGilchrist, Iain. (2009). *The Master and His Emissary: The Divided Brain and the Making of the Western World.* New Haven, CT: Yale University Press.

17. McGilchrist, Iain. (2009). *The Master and His Emissary: The Divided Brain and the Making of the Western World.* New Haven, CT: Yale University Press, p. 227.

18. This idea emerged out of personal correspondence with my editor, JoAnne O'Brien-Levin.

Chapter 7: Rethinking Language

1. Marable, Kenneth, www.msu.edu/user/marablek/whal-int.htm

2. Ibid.

3. Ibid.

4. Ibid.

5. http://www.sheldrake.org/Articles&Papers/papers/animals/parrot_telepathy.html and Sheldrake, Rupert. (1999). *Dogs That Know When Their Owners Are Coming Home.* New York, NY: Three Rivers Press, pp. 74–76.

6. I learned of this from Ryan DeMares, PhD, who played me the audiotape. DeMares received a PhD in Interspecies Communication from Union Institute, and at the time was the first person that held that particular credential (and may still be).

7. Maffi, Luisa. (Ed.). (2001). *On Biocultural Diversity: Linking Language, Knowledge and the Environment.* Washington, DC: Smithsonian Institution Press.

8. Chomsky, Noam. (1957). *Syntactic Structures.* New York, NY: Mouton.

9. Whorf, Benjamin Lee. (1956). Science and Linguistics. In *Language, Thought, and Reality.* Cambridge, MA: The MIT Press, p. 216.

10. Alford, Dan Moonhawk. (2002). Chapter Seven: The Great Whorf Hypothesis Hoax. In *The Secret Life of Language,* http://hilgart.org/enformy/dma-Chap7.htm.

11. Whorf, Benjamin Lee. (1956). Science and Linguistics. In *Language, Thought, and Reality.* Cambridge, MA: The MIT Press, p. 252.

12. Martin, Calvin Luther. (1999). *The Way of the Human Being.* New Haven, CT: Yale University Press, pp. 198–199.

13. Abram, David. (1996). *The Spell of the Sensuous.* New York, NY: Vintage Books (a division of Random House), p. 100.

14. Whorf, Benjamin Lee. (1956). The Relation of Habitual Thought and Behavior to Language. In *Language, Thought, and Reality.* Cambridge, MA: The MIT Press, p. 153.

15. http://linguistlist.org/issues/5/5-1107.html.

16. Davis, Wade. (2009). *The Wayfinders: Why Ancient Wisdom Matters in the Modern World.* Toronto: House of Anansi Press, p. 3.

17. Bohm, David. (1980). *Wholeness and the Implicate Order.* London: Routledge, p. 60.

18. Whorf, Benjamin Lee. (1956). The Relation of Habitual Thought and Behavior to Language. In *Language, Thought, and Reality.* Cambridge, MA: The MIT Press, p. 151.

19. Whorf, Benjamin Lee. (1956). The Relation of Habitual Thought and Behavior to Language. In *Language, Thought, and Reality.* Cambridge, MA: The MIT Press, p. 148.

20. Whorf, Benjamin Lee. (1956). An American Indian Model of the Universe. In *Language, Thought, and Reality.* Cambridge, MA: The MIT Press, p. 59.

21. Little Bear and Heavy Head. (2004). A Conceptual Anatomy of the Blackfoot World. *Revision, 26*(3), 31–38.

22. Little Bear and Heavy Head. (2004). A Conceptual Anatomy of the Blackfoot World. *Revision, 26*(3), 37.

23. Little Bear speaks about this in a film "Returning to Turtle Island" by Miryam Servet.

24. From a Chinese Buddhist sutra, cited in Broomfield, John. (1997). *Other Ways of Knowing: Recharting our Future with Ageless Wisdom.* Rochester, VT: Inner Traditions, p. 162.

Chapter 8: Beyond Rationality

1. Krishnamurti, J., and Bohm, David. (1999). *The Limits of Thought.* London: Routledge, p. 70.

2. Leroy told this story at the 1999 SEED dialogue, and repeated it periodically in subsequent dialogues.

3. Fromm, Erich. (1951). *The Forgotten Language: An Introduction to the Understanding of Dreams, Fairy Tales, and Myths.* New York, NY: Harry Holt and Company.

4. An interesting book on precognition is by Dossey, Larry MD. (2009). *The Science of Premonitions: How Knowing the Future Can Help Us Avoid Danger, Maximize Opportunities, and Create a Better Life.* New York, NY: The Penguin Group.

5. The Smithsonian Museum of the American Indian opened in 2004 and continues to be a magnificent living and breathing tribute to Indigenous peoples that defies the way we normally conceive of a museum—as in a monument containing artifacts from a dead past.

6. Prechtel, Martin. (1998). *Secrets of the Talking Jaguar: Memoirs from the Living Heart of a Mayan Village.* New York, NY: Jeremy Tarcher/Putnam, p. 170.

7. http://www.goodreads.com /quotes/255807-once-upon-a-time-i-dreamt-i-was-a-butterfly.

8. Some, Sobonfu. (1999). *Welcoming Spirit Home: Ancient African Teachings to Celebrate Children and Community.* Novato, CA: New World Library.

9. Ibid.

10. Arms, Suzanne. (1977). *Immaculate Deception: A New Look at Women and Childbirth in America.* New York, NY: Bantam.

11. Some, Sobonfu. (1999). *Welcoming Spirit Home: Ancient African Teachings to Celebrate Children and Community.* Novato, CA: New World Library.

12. Bennett, Hal Zina. (1993). *Zuni Fetishes: Using Native American Objects for Meditation, Reflection, and Insight.* New York, NY: Harper Collins, p. 98.

13. Leopold, Aldo. (1993). *Round River: From the Journals of Aldo Leopold* (Luna Leopold, Ed.). New York, NY: Oxford University Press, pp. 145–146.

Chapter 9: A Tale of Two Directions

1. Carson, Rachel. (1962). *Silent Spring.* Boston, MA: Houghton Mifflin, p. 277.

2. According to Honawa, Maasaw, while carrying the staff of longevity, is leaning over slightly at the end of the line, signifying that he has advanced in age.

3. Begay, David, and Maryboy, Nancy Cottrell. (1999). Nanit`a Sa` ah Naaghai Nanit`a Bik`eh Hozhoon [Living the Order: Dynamic Cosmic Process of Diné Cosmology]. *Dissertation Abstracts International, 60*(05), 1596A (UMI No. 9930321), p. 312.

Chapter 10: The Essence of Thought

1. Gibran, Kahlil. (1973). *The Prophet.* New York, NY: Alfred A. Knopf, p. 17.

2. Moulton, Janice, cited in Code, Lorraine. (1991). *What Can She Know? Feminist Theory and the Construction of Knowledge.* Ithaca, NY: Cornell University Press, pp. 23–24.

3. Kuhn, Thomas. (1996). *The Structure of Scientific Revolutions.* Chicago, IL: University of Chicago Press, p. 175.

4. Kuhn, Thomas. (1996). *The Structure of Scientific Revolutions.* Chicago, IL: University of Chicago Press.

5. Sheldrake, Rupert. (1988). *The Presence of the Past: The Memory of Nature.* Rochester, VT: Park Street Press, p. 3.

6. Blavatsky, Madame. (1888). *The Secret Doctrine: The Synthesis of Science, Religion, and Philosophy.* London: The Theosophical Publishing

Company, p. 16. Blavatsky explains "Fohat" as the "dynamic energy of Cosmic Ideation … the intelligent medium, the guiding power of all manifestation, the 'Thought Divine' … the mysterious link between Mind and Matter, the animating principle electrifying every atom into life.

7. I mean this in the generic sense. Sheldrake has a very specific usage of "formative causation" that I am not intending.

8. Wheatley, Margaret. (1999). *Leadership and the New Science: Discovering Order in a Chaotic World.* San Francisco, CA: Berrett-Koehler Publishers, pp. 122–123.

9. Vernon Masayesva, Executive Director for Black Mesa Trust, spoke about the relationship between the above and below ground waters at an extraordinary conference I attended. The conference was held on Hopi land in the spring of 2009. The name of the conference was "Braiding through Water: Weaving Traditional and Western Knowledge."

10. One of my favorite science fiction episodes from the original Star Trek TV series depicted thought as a living entity. Human beings were able to merge their energy with this thought cloud, which was capable of emotion, including love. In another Star Trek episode, a Great Crystalline Entity absorbed whole planets.

11. Steiner, Rudolph. (1968). *Theosophy: An Introduction to the Spiritual Processes in Human Life and in the Cosmos.* Pomeroy, WA: Health Research Books, p. 142–147.

12. This is a phrase often used by Nancy Maryboy during the SEED dialogues to describe learning directly from nature. She related stories of asking elders questions and receiving a reply from them—that all she needed to do was to "go to the natures" to receive information directly.

Chapter 11: To Make Thought Whole Again

1. Bohm, David. (1998). Chapter Four: The Art of Perceiving Movement. In *On Creativity* (Lee Nichol, Ed.). London: Routledge, p. 67.

2. Ibid.

3. Bohm, David. (1998). Chapter Four: The Art of Perceiving Movement. In *On Creativity* (Lee Nichol, Ed.). London: Routledge, pp. 65–66.

4. Bohm, David. (1985). *Unfolding Meaning: A Weekend of Dialogue.* London: Routledge, p. 23.

5. Ibid.

6. Bohm, David. (1985). *Unfolding Meaning: A Weekend of Dialogue*. London: Routledge, p. 24.

7. Krishnamurti, J. (1970). *The Urgency of Change*. New York, NY: Harper & Row, p. 74.

8. Krishnamurti, J. (1999). *The Limits of Thought*. London: Routledge, p. 123.

9. Lao Tsu. (1997). *Tao te Ching* (Gia-Fu Feng and Jane English, Trans.). New York, NY: Vintage Books.

10. Krishnamurti, J. (1999). *The Limits of Thought*. London: Routledge, p. 119.

11. Krishnamurti, J. (1999). *The Limits of Thought*. London: Routledge, p. 116. (Bohm made this observation in a dialogue with Krishnamurti on this topic.)

12. Krishnamurti, J. (1969). *Freedom from the Known* (Mary Lutyens, Ed.). New York, NY: Harper Collins, p. 103.

13. Campbell, Joseph. (1949). *The Hero with a Thousand Faces*. Princeton, NJ: Princeton University Press.

14. Lao Tsu. (1997). *Tao te Ching* (Gia-Fu Feng and Jane English, Trans.). New York, NY: Vintage Books (a division of Random House), p. 40.

15. Moonwalker, Tu, and O'Brien-Levin, JoAnne, with Lǎnė Saǎn Moonwalker. (2008). *Business Revolution through Ancestral Wisdom: The Circle Knowledge of the Past Comes Forward to Show Us How to Create a Practical and Ethical Process for Success, Sustainability, and True Prosperity*. Denver, CO: Outskirts Press, p. 132.

16. Berry, Thomas. (1999). *The Great Work: Our Way into the Future*. New York, NY: Bell Tower.

17. Korten, David. (2006). *The Great Turning: From Empire to Earth Community*. San Francisco, CA: Berrett Koehler; and Bloomfield, CT: Kumarian Press.

18. *Resurgence & Ecologist Magazine, March/April 2013*(277), 35–37.

19. Isha Upanishad, http://www.osholeela.com/poetry/isa.html.

Chapter 12: To Think Without Separation

1. Berry, Thomas. (1999). *The Great Work: Our Way into the Future.* New York, NY: Bell Tower, p. 72.

2. Benyus, Janine. (1997). *Biomimicry: Innovation Inspired by Nature.* New York, NY: Harper Collins, p. 97.

3. Berry, Thomas. (1999). *The Great Work: Our Way into the Future.* New York, NY: Bell Tower, pp. 76–77.

4. Thomas Aquinas cited in Berry, Thomas. (1999). *The Great Work: Our Way into the Future.* New York, NY: Bell Tower, p. 77.

5. Morgan, Faith. "The Power of Community: How Cuba Survived Peak Oil." This film was made after Cuba went through an oil embargo from the Soviet Union, which initially caused chaos, but eventually resulted in the community successfully returning to sustainable agriculture in small plots, often on city roofs. The status of farmers was greatly elevated during this period and demonstrable health improvements occurred because of high quality food.

6. My high school friend: Don Schwartz.

7. Bateson, Nora. (2011). *An Ecology of Mind: A Daughter's Portrait of Gregory Bateson.* Oley, PA: Bullfrog Films.

8. Johann Wolfgang von Goethe, cited in Spretnak, Charlene. (1999). *Resurgence of the Real: Body, Nature, and Place in a Hypermodern World.* London: Routledge, p. 138.

9. http://scienceworld.wolfram.com/biography/Kelvin.html.

10. Bohm, David. (1980). *Wholeness and the Implicate Order.* London: Routledge, p. xvii.

11. Tolman, Deborah L., and Brydon-Miller, Mary. (2001). *From Subjects to Subjectivities: A Handbook of Interpretive and Participatory Methods.* New York, NY: New York University Press.

12. This story was told to me by Dan Wildcat, professor at Haskell University. Dan worked closely with Vine Deloria Jr. and coauthored the book *Power and Place* with him.

Chapter 13: Rethinking the "Dismal Science"

1. Klein, Julie Thompson. (1990). *Interdisciplinarity: History, Theory & Practice.* Detroit, MI: Wayne State University Press, p. 22.

2. Daly, Herman, and Cobb, John. (1989). *For the Common Good: Redirecting the Economy toward Community, the Environment, and a Sustainable Future.* Boston, MA: Beacon Press, p. 124.

3. Daly, Herman, and Cobb, John. (1989). *For the Common Good: Redirecting the Economy toward Community, the Environment, and a Sustainable Future.* Boston, MA: Beacon Press, p. 32.

4. Schumacher, E. F. (1973). *Small Is Beautiful: Economics as if People Mattered.* New York, NY: Harper Perennial, p. 43.

5. Spretnak, Charlene. (1999). *Resurgence of the Real: Body, Nature, and Place in a Hypermodern World.* London: Routledge, p. 142.

6. Schumacher, E. F. (1973). *Small Is Beautiful: Economics as if People Mattered.* New York, NY: Harper Perennial, p. 57.

7. Daly, Herman, and Cobb, John. (1989). *For the Common Good: Redirecting the Economy toward Community, the Environment, and a Sustainable Future.* Boston, MA: Beacon Press, p. 30.

8. Daly, Herman, and Cobb, John. (1989). *For the Common Good: Redirecting the Economy toward Community, the Environment, and a Sustainable Future.* Boston, MA: Beacon Press, p. 164.

9. There is a subdiscipline emerging today called behavioral economics, that is at long last reversing this trend and realizing that people do not really act only in self-serving, rational ways.

10. Daly, Herman, and Cobb, John. (1989). *For the Common Good: Redirecting the Economy toward Community, the Environment, and a Sustainable Future.* Boston, MA: Beacon Press, p. 86.

11. Daly, Herman, and Cobb, John. (1989). *For the Common Good: Redirecting the Economy toward Community, the Environment, and a Sustainable Future.* Boston, MA: Beacon Press, p. 47.

12. Gorman, James. (2013, February 7). Bacteria Found Deep under Ice, Scientists say, Opening New Antarctic World. *New York Times,* p. A18.

13. Powell, Diane Hennacy MD. (2009). *The ESP Enigma: The Scientific Case for Psychic Phenomena.* New York, NY: Walker Publishing Company, p. 181.

14. Wheatley, Margaret. (1999). *Leadership and the New Science: Discovering Order in a Chaotic World.* San Francisco, CA: Berret-Koehler, p. 98.

15. Sheldrake, Rupert. (2012). *The Presence of the Past: Morphic Resonance & the Memory of Nature.* Rochester, VT: Park Street Press, p. 64.

16. Da Vinci, Leonardo. (1941). *The Notebooks of Leonardo da Vinci* (Arranged, Rendered into English, and Introduced by Edward McCurdy). Garden City, NY: Garden City Publishing, p. 666–667.

17. Da Vinci, Leonardo. (1941). *The Notebooks of Leonardo da Vinci* (Arranged, Rendered into English, and Introduced by Edward McCurdy). Garden City, NY: Garden City Publishing, p. 654.

Chapter 14: Toward an Original Economics

1. Daly, Herman, and Cobb, John. (1989). *For the Common Good: Redirecting the Economy toward Community, the Environment, and a Sustainable Future.* Boston, MA: Beacon Press, p. 139.

2. Daly, Herman, and Cobb, John. (1989). *For the Common Good: Redirecting the Economy toward Community, the Environment, and a Sustainable Future.* Boston, MA: Beacon Press, p. 138.

3. I once had a conversation with Larry Dossey about how the term "integrative medicine" was designed to subsume alternatives under this name, and he remarked to me "Oh, you noticed that too."

4. Daly, Herman, and Cobb, John. (1989). *For the Common Good: Redirecting the Economy toward Community, the Environment, and a Sustainable Future.* Boston, MA: Beacon Press, p. 138.

5. It is defined this way in every issue of the journal *Ecological Economics* under "Aims and Scope."

6. Daly, Herman, and Cobb, John. (1989). *For the Common Good: Redirecting the Economy toward Community, the Environment, and a Sustainable Future.* Boston, MA: Beacon Press, p. 139.

7. Eisler, Riane. (2007). *The Real Wealth of Nations. Creating a Caring Economics.* San Francisco, CA: Berrett-Koehler, p. 28.

8. Ibid.

9. Campbell, Joseph. The complete four volume set, published by Viking Press, New York, includes: Vol. 1 (Primitive Mythology); Vol. II (Oriental Mythology); Vol. III (Occidental Mythology); and Vol. IV (Creative Mythology).

10. http://www.heartmath.org/research/science-of-the-heart/head-heart-interactions.html?submenuheader=3.

11. Taylor, Jill Bolte. (2006). *My Stroke of Insight: A Brain Scientist's Personal Journey.* New York, NY: Penguin.

12. http://www.youtube.com/watch?v=gutCn0oPbS.

13. Eisenstein, Charles. (2011). *Sacred Economics: Money, Gift, and Society in the Age of Transition.* Berkeley, CA: Evolver, http://sacred-economics.com/read-online/.

14. Richard Nixon is a hero to many in Taos Pueblo because he returned the Blue Lakes Lands of Taos Mountain to the tribe in perpetuity, ensuring that tribal members would continue to drink pure, fresh water for as long as they live.

15. DDT, while not used for agricultural purposes in Mexico, is still a considerable public health problem. See http://www.ncbi.nlm.nih.gov/pmc/articles/PMC1469381/

16. Cunningham, Storm. (2002). *The Restoration Economy: The Greatest New Growth Frontier.* San Francisco, CA: Berrett-Koehler, p. 2.

17. Ibid., p. 7.

18. Ibid., p. 304.

19. Ibid., pp. 80–81.

20. A traditional Tanoan Indian Story told by Cajete, Gregory. (1994). *Look to the Mountain: An Ecology of Indigenous Education.* Skyland, NC: Kivaki Press, pp. 125–128.

Chapter 15: Education as Renewal

1. Benyus, Janine. (1997). *Biomimicry: Innovation Inspired by Nature.* New York, NY: Harper Collins, pp. 254–278.

2. Holdrege, Craig. (2013). *Thinking Like a Plant: A Living Science for Life.* Great Barrington, MA: Lindisfarne, p. 164.

Chapter 16: Childhood and Education

1. My writing on Waldorf education was influenced from discussion with Waldorf educators, especially Richard Ailes and Susan Junge.

2. Abram, David. (2010). *Becoming Animal: An Earthly Cosmology.* New York, NY: Pantheon (division of Random House), pp. 38–39.

3. Louv, Richard. (2008). *Last Child in the Woods: Saving Our Children from Nature-Deficit Disorder.* Chapel Hill, NC: Algonquin Books.

4. Excerpted from Linda Hogan quotation. The full quotation is at the beginning of Part Two (Departure) of this book.

5. http://www.ncbi.nlm.nih.gov/pmc/articles/PMC153612
/ and https://answersingenesis.org/genetics/dna-similarities
/humanchimp-dna-similarity-continues-to-decrease-counting-indels/.

There is some controversy over these estimates of human-chimpanzee DNA correlation, and undoubtedly the estimates will change over time. The main point, however, remains the same: we are closely related to the rest of life.

6. Watts, Alan. (1966). *The Book on the Taboo Against Knowing Who You Are.* New York, NY: Random House, p. 40.

Chapter 17: Higher Education

1. The best discussion I have found of this trend is in Wilshire, Bruce. (1990). *The Moral Collapse of the University.* Albany, NY: State University of New York Press.

2. Hawken, Paul. (2007). *Blessed Unrest: How the Largest Movement in the World Came Into Being and Why No One Saw It Coming.* New York, NY: Penguin.

3. Whitehead, Alfred North. (1925). *Science and the Modern World.* New York, NY: Free Press.

4. _____. (1938). *Modes of Thought.* New York, NY: Macmillan, p. 11.

5. Benyus, Janine. (1997). *Biomimicry: Innovation Inspired by Nature.* New York, NY: Harper Collins, p. 260.

6. Klein, Julie Thompson. (1990). *Interdisciplinarity: History, Theory & Practice.* Detroit, MI: Wayne State University Press, pp. 22–24. According to Klein, the modern concept of interdisciplinarity began in the twentieth century shortly on the heels of the formation of the first modern disciplines (history in 1884, economics in 1885, political science in 1903, sociology in 1905 et al.) and was linked with the progressive vision of Alfred North Whitehead and John Dewey. Klein does concede that the origins of interdisciplinarity can be seen in the Greek *enkudlios paedeia* and the Roman *orbis doctrinae* (programs that reflect the belief that a well-rounded person is one who has surveyed the breath of the disciplines). Klein also notes that William Von Humboldt's concept of "universal education" began in the nineteenth century, but this was basically an anomaly at the time.

7. Wilshire, Bruce. (1990). *The Moral Collapse of the University.* Albany, NY: State University of New York Press.

8. Giles, James. (2000, Spring/Summer). The End of Philosophy. *Cross Currents: The Journal of Addiction and Mental Health,* 68–76. Also see Wilshire, Bruce. (1990). *The Moral Collapse of the University.* Albany, NY: State University of New York Press.

9. In the initial drafts of Eisenhower's famous speech in which he mentions the "military-industrial complex," he included the word "academic." The full phrase was the "military- industrial-academic complex." He deleted the word "academic" before delivering his speech. See Giroux, Henry A. (2007). *The University in Chains: Confronting the Military-Industrial-Academic Complex.* Boulder, CO: Paradigm, pp. 14–15.

10. Taylor, Mark C. (2009, April 23). End the University as We Know It. *New York Times.*

11. Often quoted, did not necessarily originate with Buckminster Fuller. Fuller also used to like to say: "You can never learn less, you can only learn more. The reason I know so much is because I have made so many mistakes." And he once advocated that the school children who make the most mistakes should get the best grade (because that means they learned the most).

Chapter 18: A New (and Ancient) Vision

1. Steiner, Rudolph. (1995). *Intuitive Thinking as a Spiritual Path: A Philosophy of Freedom.* Great Barrington, MA: Anthroposophic Press, p. 25.

2. Targ, Russell, and Katra, Jane. (1998). *Miracles of Mind: Exploring Nonlocal Consciousness and Spiritual Healing.* Novato, CA: New World Library, p. 273.

3. My thoughts on this subject are informed by several elders, including Little Bear, Tobasonakwut Kinew, and Grandfather Leon Secatero. Any true Indigenous elder I have known are incredibly observant of natural phenomena.

Chapter 19: A Vision for Original Thinking

1. Meade, Michael. (2012). *Fate and Destiny: The Two Agreements of the Soul.* Seattle, WA: Mosaic Multicultural Foundation, p. 33.

2. Whitehead, Alfred North. (1925). *Science and the Modern World.* New York, NY: Free Press.

3. Benyus, Janine. (1997). *Biomimicry: Innovation Inspired by Nature.* New York, NY: Harper Collins, p. 99.

4. Martin Chesler, private conversation.

5. Prechtel, Martin. (2012). *The Unlikely Peace at Cuchumaquic: Parallel Lives of People as Plants: Keeping the Seeds Alive.* Berkeley, CA: North Atlantic Books, p. xvi.

INDEX

Hegel, Georg Wilhelm Friedrich, 177
Henderson, Hazel, 189, 191
Henderson, Sakej, 95, 103–4
Hero's journey, 160
Historical consciousness, 6–7, 100
Hobbes, Thomas, 75–76
Hogan, Linda, 56, 115
Honawa, 126
Hopi, 40, 105, 106–7, 126–27, 147,
    249
House, Donna, 78–79, 115
Hubble telescope, 23
Hubris, 19, 165
Human beings
    as animals, 64–65
    attributes of, 63–64, 69–70
    dependence of, upon the whole, 130,
        166
    evolution of, 29
    imagined superiority of, 69–70, 76,
        91–92, 164–66
    instincts in, 64
    interconnection of, 224
    as observers, 20
    potential of, 62–63
    purpose of, 119, 211, 212
    relationship of nature and, 128–30,
        165–66
    survival of, 127, 167–68, 265
Human culture
    emergence of, 122–23
    refinement of, 122
Humanness
    ancient Greek view of, 72
    birth and, 117–19
    community and, 123
    meaning of, 61–62
    new (original) assumptions about,
        242–43
    non-Western view of, 125
    paradox of, 130–33
    rational thought and, 62, 63, 71,
        72–73, 78, 89

tacit modern assumptions about, 68
technology and, 85–86
totality of, 63–64
Western view of, 72–73, 88–90,
    109–10, 125, 168

I
Implicate order, 33
Impressionism, 49
Inca, 40
Indigenous consciousness
    origin and, 3–4
    quantum theory and, 24–25
    time and, 9–11, 101
    Western vs., 3
Indigenous languages
    drawbacks of, 104
    passing of time expressed by, 105–7
    richness of, 103–4
    singing and chanting, 98
Individuality
    meaning of, 219–20
    in Western society, 118
Industrial revolution, 22–23, 161, 230
Information
    stories vs., 255
    wisdom vs., 264
Inspiration, 65, 81, 112
Instinct, 64, 65–67, 70
Integral consciousness, 25, 37
Intelligence, 91–92
Interdisciplinary studies, 231
Intuition, 76–77, 82–83, 88, 90, 193
Inuit people, 96
Involution, 27, 28

J
James Bay Hydroelectric project, 239
Janus, 17
Japan, sense of self in, 225
Jeans, James, 2, 146
Jesus, 27, 240
Josephson, Brian, 79

ACKNOWLEDGMENTS

First, I thank all the ancestors who shaped this land and shaped my consciousness to the extent that I am able to hear their messages. I am forever grateful to a select group of ancestors whom I was blessed to know and love while they were alive, including Grandfather Leon Secatero, Dan Moonhawk Alford, Tobasonakwut Kinew, Paula Gunn Allen, Patricia Clark Smith, Kent O'Gormley, and my parents Joan and Juan Paricio Parry. I am also indebted to those who came before me whom I never met but whose work made a lasting impression. This group would include David Bohm, who did much to create a fertile ground for dialogue and the coming together of Western consciousness and Indigenous consciousness; J. Krishnamurti, William James, John Dewey, Gregory Bateson, and Buckminster Fuller for their ideas on education, thinking, life, and what it means to be human; Benjamin Lee Whorf and Owen Barfield for their ideas on language and the origin of words; Sri Aurobindo, Jean Gebser, and Madame Blavatsky for their contribution in articulating the emerging evolution of consciousness; E. F. Schumacher for his ideas on restoring the original meaning of economics; and Rachel Carson, Barry Commoner, and Thomas Berry for their wisdom on ecology.

I give great thanks to my editor, JoAnne O'Brien-Levin, who is literally the editor of my dreams, for I dreamed of her necessary involvement at a certain stage. JoAnne is not just an editor but more of a "thinking partner" who is a skillful listener and coach. I also thank Vesela Simic, Howard McConeghey, Lee Nichol, David Abram, Robert C. Koehler, Anthony Dellaflora, Jeffrey Paine, Jean Houston, and Naomi Rose for review of the manuscript at various stages. I thank Richard Ailes and his wife Susan for their thoughts on Waldorf education and Rudolph Steiner, and Robert

Schimelpfenig for our discussions on Socrates. I give thanks for my agent Bill Gladstone, and for Leslie Larson, Tim McKee, Johanna Wallner, Julia Kent, Kim Westerman, and all the folks at North Atlantic Books who helped to finalize the submitted manuscript for publication and to get the book distributed. And I thank my wonderful publicity team, including Jill Mangino, Colby DeVitt, Kennedy Braden, and webmaster Canton Becker for getting the message out.

In truth, I am indebted to so many people that the work as a whole could be thought of as one giant acknowledgments section. I feel similarly to Martin Prechtel about this sort of thing, who recently attributed his first book to *káslimáal,* the Mayan word for "mutual indebtedness" to all of "life's things to all other things before and after."[5] What he says is accurate; nothing happens in isolation. We are not only all connected; we can also learn from everyone. As James O'Dea, one of my mentors, likes to say, "Everywhere we turn is the face of our teacher."

That being said, there are certain elders, peers, and loved ones that I wish to specifically acknowledge, as without them there would have been no book or even a thought of it. First and foremost, I thank both my wise, patient, and loving wife, Tomoko, and also Leroy Little Bear for his wise and skillful moderation of the Language of Spirit annual conferences from 1999 through 2011. I thank not just Leroy, but all the participants in the dialogue circle, both inner and outer circles, for it is largely through the process of collective knowledge construction that occurs in dialogue that I am able to impart what I have come to know for the benefit of those that come after me.

Amongst the many people who participated in these dialogues, some giants have departed, including the aforementioned Moonhawk, Grandfather Leon Secatero, Tobasonakwut Kinew, and Paula Gunn Allen. But there are also many wonderful people from these dialogues that I am blessed to know today, including Lee Nichol, Alfonso Montuori, Nancy Maryboy, David Begay, James O'Dea, Amethyst First Rider, Matthew Bronson, Gregory Cajete, Joseph Rael, Linda Hogan, Donna House, David Abram, Polly Walker, Dennis Martinez, Angelita Borbon, Woman Stands Shining (Pat McCabe), David Peat, Phillip Duran, Rose

Von Thater-Braun, MJ Zimmerman, Lloyd Pinkham, Ashok Gangadean, Edgar Mitchell, James Ryan, Marilyn Mandala Schlitz, Brian Josephson, Brian Swimme, Fred Alan Wolf, Amit Goswami, William Tiller, Phillip Sakimoto, Beverly Rubik, Norm Sheehan, Jill and Gladys Milroy, Tjilaminu Mia, Grandmother Mona Polacca, Dan Longboat, and Leslie Gray.

This book would not be possible were it not for a host of other experiences I had with Source for Educational Empowerment and community Dialogue (SEED) of Albuquerque, New Mexico, United States, the organization I founded in 1996 and served until retiring at the end of 2012. In certain respects, SEED is a child of mine, and I wish her well as she now ventures out in a different direction. But mostly, SEED was my greatest teacher, for it was there that I met so many wise and wonderful people who helped me develop and clarify my own thinking. I thank all the past members of the Board of Trustees whom I served alongside, as well as the staff, volunteers, and supporters, each and every one of them. They sustained not just the organization; they also sustained me over the years. I particularly thank Shawn Secatero, son of Grandfather Leon Secatero, who served as president while I was writing the bulk of this book.

I thank the many elders and scholars I have known, including six members of the International Council of 13 Indigenous Grandmothers that participated at the 2012 Wisdom from the Origins conference (Grandmothers' Flordemayo, Agnes Baker Pilgrim, Beatrice and Rita Long Visitor Holy Dance, Margaret Behan, and Mona Polacca); Don Alejandro Cerilo Perez Oxlaj, Sequoyah Trueblood, Black Eagle Sun, Sakej Youngblood Henderson, Larry Littlebird, Felipe Ortega, Rangamarie Rose Pere, Gregg Braden, Dan Wildcat, Patricio Dominguez, Jose Villa, and Harlan McKosato. And I thank professor Tamara M. Green, who I took a class from at NYU over thirty years ago that changed my life. The class was called "Magic, Witchcraft, Alchemy, The Occult, and the Emergence of Modern Science."

I thank Alvira Yazzie, Leon's eldest daughter, Orlando Secatero, his eldest son, and Shawn Secatero, the youngest son, and Renee Secatero, for their friendship and ongoing connection to their father and their sacred lands. I thank all the people who have permitted me to participate in ceremony, including Woman Stands Shining, Stephen Bluehorse, Tobasonakwut Kinew, Michael Two Bears Andrews, the entire Secatero family, the venerable Grandmother Delhia, (who speaks to me in Navajo and somehow I manage to occasionally understand), and to all the folks who are blessed to gather every year at the Canoncito sacred grounds, including Will, Susan, Gail, Bonnie, Thelma, Florence, Juana, Lou, Harriette, Mr. Frank, Elsie, JJ, Leland, JR, Smudge, Mr. Brady, Gomo, and others.

For their love and support, I thank my brother and sister, Neil Parry and Linda Paricio, and their spouses Diane and Brian; I thank all my loving aunts and uncles, Ray and Jean Paricio, Ado and Pity, and all their children; my Spanish and Basque relatives Pilar Paricio Perez, Cuca, Eugenio, Paulino and Severina, Carlos Mairal, and all the people from the village of Ceresa in the High Pyrenees.

There are so many others I could thank for their counsel, wisdom, and practical assistance or support of one kind or another, but amongst the most influential are Alan Hutner and Elizabeth Rose, Joyce Anastasia, Eleonore Caracciolo, Neal Sutton, Jacqueline Kane Babbitt, Carole Hart, Victoria Hanchin, Ursula Daniels, Christina Tonan, Dr. Stephen Weiss, Paula Bandy, Carole Rominger, Pravin Agarwal, Ditta Khalsa, Pele Rouge Chadhima, Lana Holmes, Rumi Dover, Michele Rozbitsky, Szuson Wong, Doug Cohen, Peggy Hawkins, Miryam Servet, Connie, Matthew, and Mary of Foundation for Global Humanity, Carlos Villatoro, Nicki La Rosa, Joseph Subbiondo, Dennis Cole, David Vogel, Marietta Leis, Herman Daly, Satish Kumar, Tony Abeyta, Joe Cajero, Clare Villa, Aryeh Swisa, Jeff Aitken, Robin Globus, Barbara Kohl, Jacqueline Trent, Doreen Goodlin, Diane Oberlin, Robert and Linda Lupowitz, Michael Logghe, Art Goodtimes, Bob XI, Halima Christy, Kimberley A. Jackson, Richard and Patricia Dobrin, Martin Chesler, Peter Silverman, Elizabeth Cogburn, Genevieve Chavez Mitchell, Lisa Tannebaum, Dee Jones Bartholomew, Paula Amar Schwartz, Marie Mound, Maxine Freed, Alorah Lavender, Ralph Wolff, Galen Knight, Kathrin Morris, Marlene Simon, David Barnes, David Biggs, Cheryl Skinner, Jacqueline Wasilewski, Camille

Brito, Jim Hartshorn, Barbara Gordon, Marilyn Jesmain, Jeanette Rattner, Karen Kudebeh, Laurelyn Baker, Reuben Medina, Joe Garcia, Ramona Emerson and Kelly Byars, Zia Pueblo, Asha Lela, Medea Eder, Karen Dallett, Roberto Gonzalez-Plaza, Dyanne Strongbow, James Jereb, Diego Mulligan, Dr. Swedberg, Carolyn Genet Martin, Junko Endo, Dr. Lucy Ferguson, Dr. James Slaman, Yolanda Chavez, Sunrise, Haru, Satchmo, and Adobe.

And finally, I am indebted to the wisdom of nature and the special places in New Mexico that have nourished me: the Sandia and Taos mountains; the Rio Grande; the clouds, canyons, gorges, and caves; the light and soil of this precious land that always inspires me. In keeping with this gift, I give thanks to the Helene Wurlitzer Foundation and their executive director, Michael Knight, a sweet soul, for providing me a pair of fellowships where I could begin and nearly end this manuscript in Taos, NM, a beautiful and magical setting of grandfather cottonwood trees, ravens, magpies, and a blanket of snow. I am also grateful to the Wood Family of Apple Hill Farm in Arroyo Seco just north of Taos, where I stayed in beauty for a week in the crisp early spring and finished the manuscript.

ircle for
riginal Thinking

The vision of the Circle for Original Thinking is to restore thinking to its origins and full spectrum—in tune with the way Nature thinks—in order to create a more whole and just world for all our relations on our beautiful planet. We are assisted in this process by remembering that the pulse of the human heart is one with the heart of Nature and that all of Nature is alive and aware, our deep kin, and worthy of respect. Thinking then becomes inclusive—incorporating intuition, feeling, expanded sensory awareness, and a larger sense of "what wants to happen."

The Circle for Original Thinking is an inclusive, grass-roots organization that honors the ancient and modern, feminine and masculine, Indigenous, Eastern and Western philosophies through their deep, interconnected roots. As we listen and think from the origins, we shift and shape thinking towards wisdom and wholeness—and seek to bring this heart-centered wisdom into contemporary society. We do this by conducting wholistic scientific research in the spirit of the ancients who learned from the patterned integrity in Nature; publishing creative papers, books, poetry; and through organizing intercultural dialogue circles, conferences, ceremonies, concerts, and other multi-media events.

We invite your participation and support in this inspired and comprehensive vision. Membership is open to all.

To join the Circle for Original Thinking, or for more information, visit www.originalthinking.us

You can also write to us at:
Circle for Original Thinking
12231 Academy NE #307–244
Albuquerque, NM 87111

ABOUT THE AUTHOR

Photo: Kyle Zimmerman

Glenn Aparicio Parry, PhD, also given the name *Kizhe Naabe* (Ojibwe for Kind-Hearted Man), is a writer, educator, international speaker, entrepreneur, and visionary whose lifelong passion is to reform thinking and education into a coherent, cohesive whole. The founder and past president of the SEED Institute, Parry is currently the president of the think tank: *The Circle for Original Thinking*. He earned his BA in Psychology from Allegheny College and went on to earn both his MA in East-West Psychology and his PhD in Humanities with a concentration in Transformative Learning from the California Institute of Integral Studies. Parry organized and participated in the groundbreaking Language of Spirit Conferences from 1999–2011 that brought together Native and Western scientists in dialogue, moderated by Leroy Little Bear. He is a member of the Institute of Noetic Sciences and the Theosophical Society as part of a lifelong interest in bridging the arts and sciences. An avid outdoorsman, he now makes his home in the foothills of the Sandia Mountains in Albuquerque, New Mexico, with his wife Tomoko, dog Sunrise, and cat Haru (Japanese for spring).